Collins INTERNATIONAL PRIMARY SCIENCE

Teacher's Guide 1

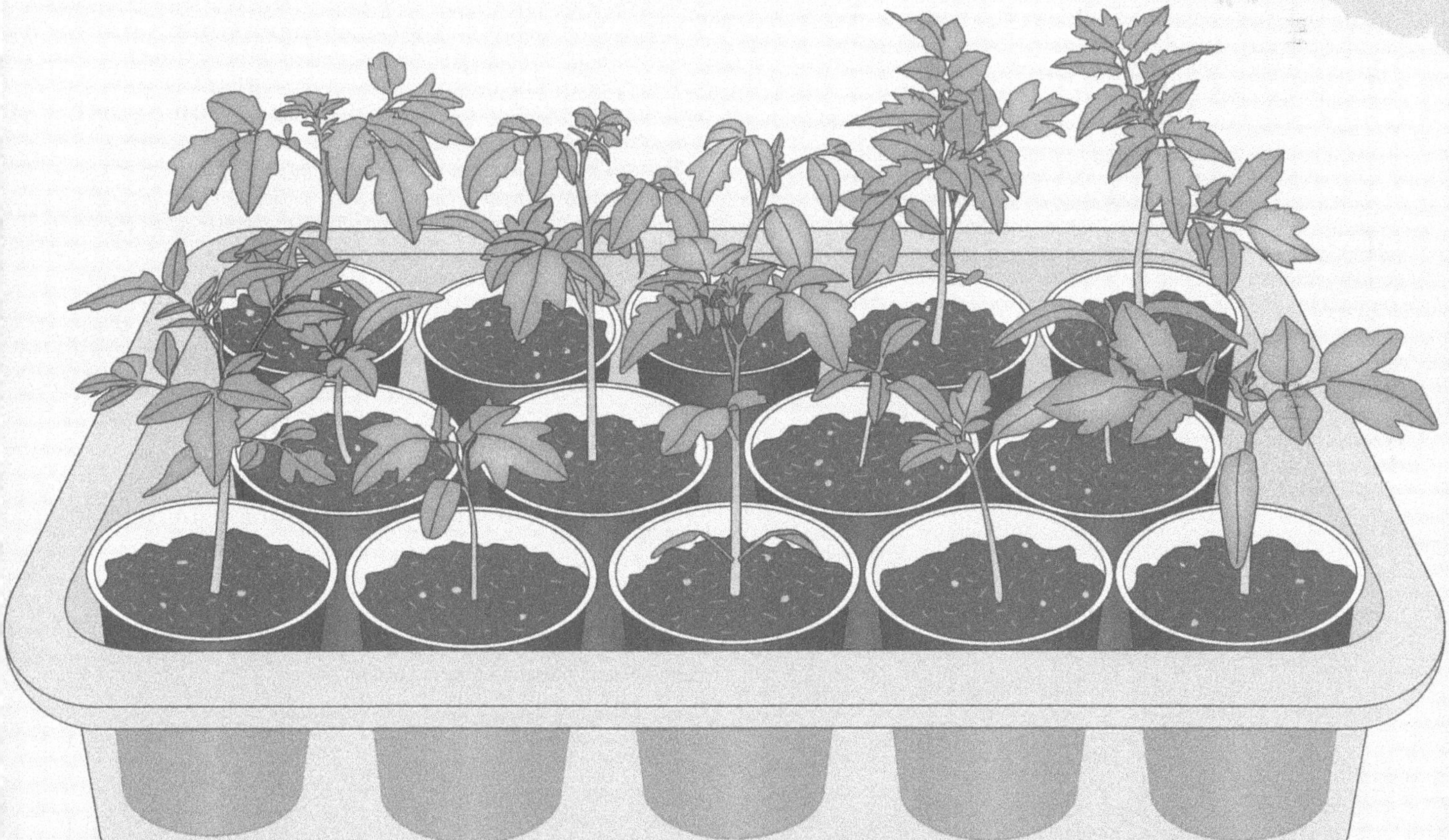

William Collins' dream of knowledge for all began with the publication of his first book in 1819.
A self-educated mill worker, he not only enriched millions of lives, but also founded a flourishing publishing house. Today, staying true to this spirit, Collins books are packed with inspiration, innovation and practical expertise. They place you at the centre of a world of possibility and give you exactly what you need to explore it.
Collins. Freedom to teach.

Published by Collins
An imprint of
HarperCollins*Publishers* Ltd.
The News Building
1 London Bridge Street
London
SE1 9GF

HarperCollins*Publishers*
Macken House
39/40 Mayor Street Upper
Dublin 1
DO1 C9W8
Ireland

Browse the complete Collins catalogue at
www.collins.co.uk

© HarperCollins*Publishers* Limited 2021

11

ISBN: 978-0-00-836899-9

Second edition

Contributing authors: Phillipa Skillicorn, Tracy Wiles, Karen Morrison, Tracey Baxter, Sunetra Berry, Pat Dower, Helen Harden, Pauline Hannigan, Anita Loughrey, Emily Miller, Jonathan Miller, Anne Pilling, Pete Robinson.

Any educational institution that has purchased one copy of this publication may make unlimited duplicate copies for use exclusively within that institution. Permission does not extend to reproduction, storage within a retrieval system, or transmittal in any form or by any means, electronic, mechanical, photocopying, recording or otherwise, of duplicate copies for loaning, renting or selling to any other institution without the permission of the Publisher.

Without limiting the exclusive rights of any author, contributor or the publisher of this publication, any unauthorised use of this publication to train generative artificial intelligence (AI) technologies is expressly prohibited. HarperCollins also exercise their rights under Article 4(3) of the Digital Single Market Directive 2019/790 and expressly reserve this publication from the text and data mining exception.

British Library Cataloguing in Publication Data
A Catalogue record for this publication is available from the British Library.

Commissioning editor: Joanna Ramsay
Product manager: Letitia Luff
Development editor: Karen Williams
Project manager: 2Hoots Publishing Services Ltd
Proofreader: Caroline Low
Cover designer: Gordon MacGilp
Cover illustrator: Ann Paganuzzi
Image researcher: Emily Hooton
Illustrators: Beehive Illustration (John Batten, Moreno Chiacchiera, Phil Garner, Kevin Hopgood, Tamara Joubert, Simon Rumble, Jorge Santillan, Matt Ward); Graham-Cameron Illustration (Sue Woollatt)
Internal design and typesetting: Ken Vail Graphic Design Ltd
Production controller: Lyndsey Rogers
Printed and bound by: Ashford Colour Ltd

Cambridge International copyright material in this publication is reproduced under licence and remains the intellectual property of Cambridge Assessment International Education.
Third-party websites, publications and resources referred to in this publication have not been endorsed by Cambridge Assessment International Education.

The assessment-style questions and sample answers used in the Topic quiz sheets have been written by the authors. These and references to assessment and/or assessment preparation are the publisher's interpretation of the curriculum framework requirements and may not fully reflect the approach of Cambridge Assessment International Education.

With thanks to the following teachers and schools for reviewing materials in development: Preeti Roychoudhury, Sharmila Majumdar and Sujata Ahuja, Calcutta International School; Hawar International School; Melissa Brobst, International School Budapest; Rafaella Alexandrou, Diana Dajani, Sophia Ashiotou and Adrienne Enotiadou, Pascal Primary School Lefkosia; Niki Tzorzis, Pascal Primary School Lemesos; Vijayalakshmi Chillarige, Manthan International School; Taman Rama Intercultural School.

Acknowledgements
The publishers wish to thank the following for permission to reproduce photographs.
Every effort has been made to trace copyright holders and to obtain their permission for the use of copyright materials. The publishers will gladly receive any information enabling them to rectify any error or omission at the first opportunity.

p103t jadimages/Shutterstock, p103tc Odua Images/Shutterstock, p103bc azhuvalappil/Shutterstock, p103b Hamady/Shutterstock, p116tl Fotovika/Shutterstock, p116tr Bragin Alexey/Shutterstock, p116tcl Johannes Kornelius/Shutterstock, p116tcr Richard Peterson/Shutterstock, p116bcl aopsan/Shutterstock, p116bcr & 124tcl photosync/Shutterstock, p116bl kuppa/Shutterstock, p116br Fribus Ekaterina/Shutterstock, p124tl Ruslan Semichev/Shutterstock, p124tr Gjermund/Shutterstock, p124tcr Julia Ivantsova/Shutterstock, p124bcl Mikhail Kniazev/Shutterstock, p124bcr r.classen/Shutterstock, p124bl studiovin/Shutterstock, p134t Africa Studio/Shutterstock, p134tc K Serhii/Shutterstock, p134c Sergey Novikov/Shutterstock, p134bc travellight/Shutterstock, p134b Pawel Nawrot/Shutterstock, p135tl Oleksandr_Delyk/Shutterstock, p135tr hobbit/Shutterstock, p135tcl livingpitty/Shutterstock, p135tcr Andrey Arkusha/Shutterstock, p135bcl Hurst Photo/Shutterstock, p135bcr Photoongraphy/Shutterstock, p135bl FeellFree/Shutterstock, p135br Ljupco Smokovski/Shutterstock, p152tl nikkytok/Shutterstock, p152tr KJBevan/Shutterstock, p152tcl Eric Boucher/Shutterstock, p152tcr owen1978/Shutterstock, p152bcr holbox/Shutterstock, p152bl Aleksandr Bryliaev/Shutterstock, 152br ayzek/Shutterstock, p138l magicoven/Shutterstock, p138cl Karkas/Shutterstock, p138cr Crepesoles/Shutterstock, p138r Pavlo Loushkin/Shutterstock.

This book contains FSC™ certified paper and other controlled sources to ensure responsible forest management.

For more information visit: www.harpercollins.co.uk/green

Contents

Introduction	v
Teacher's Guide	vi
Student's Book	viii
Workbook and Digital resources	x
Cambridge Global Perspectives™	xi
Assessment in Cambridge Primary Science	xii
Learning objectives matching grid	xiv
Lesson plans	1

Topic 1 Plants

1.1	All about Science	2
1.2	Is it alive?	4
1.3	Plants and animals are living things	6
1.4	Things that have never been alive	8
1.5	Parts of a plant	10
1.6	What do plants need to survive?	12
	Consolidation and Topic quiz answers	14
	Student's Book answers	15

Topic 2 Humans and other animals

2.1	Parts of the human body	16
2.2	Our senses	18
2.3	Using our senses	20
2.4	Users of science	22
2.5	What do animals need to survive?	24
2.6	Humans are similar	26
2.7	Humans are different	28
	Consolidation and Topic quiz answers	30
	Student's Book answers	31

Topic 3 Materials

3.1	Similar or different?	32
3.2	Properties of materials	34
3.3	More properties	36
3.4	What material is it?	38
3.5	More materials	40
3.6	Sorting materials	42
3.7	Making smaller groups	44

3.8	Materials can change shape	46
3.9	Squashing and bending	48
3.10	Stretching and twisting	50
3.11	Uses of science	52
	Consolidation and Topic quiz answers	54
	Student's Book answers	55

Topic 4 Forces and sound

4.1	Thinking and working scientifically	56
4.2	Movement	58
4.3	Pushing and pulling	60
4.4	Pushes and pulls	62
4.5	Floating and sinking	64
4.6	Listen carefully	66
4.7	What made that sound?	68
4.8	Loud and quiet sounds	70
4.9	Sound and distance	72
	Consolidation and Topic quiz answers	74
	Student's Book answers	75

Topic 5 Electricity and magnetism

5.1	What things need electricity?	76
5.2	Exploring magnets	78
5.3	History of science	80
	Consolidation and Topic quiz answers	82
	Student's Book answers	83

Topic 6 Earth and Space

6.1	Clean water investigation	84
6.2	Our planet Earth	86
6.3	Science and the environment	88
6.4	What is land made of?	90
6.5	The Sun	92
	Consolidation and Topic quiz answers	94
	Student's Book answers	95

Photocopy Masters 96

Topic quizzes 153

Global Perspectives 167

Introduction

About *Collins International Primary Science*

Collins International Primary Science is specifically written to fully meet the requirements of the Cambridge Primary Science curriculum framework, and the material has been carefully developed to meet the needs of primary science students and teachers in a range of international contexts.

Content is organised according to the four main strands: Biology, Chemistry, Physics and Earth and Space, and the skills detailed under the Thinking and Working Scientifically strand are introduced and taught in the context of those areas.

All course materials make use of the fully integrated digital resources. For example, video clips and slideshows allow students the opportunity to view at first-hand examples of habitats, plants and animals they may not be familiar with from their own country. The interactive activities provide a valuable teaching resource that will engage the students and consolidate learning.

Components of the course

For each of Stages 1 to 6 as detailed in the Cambridge Primary Science curriculum framework, we offer:

- a full colour, highly illustrated and photograph-rich Student's Book
- a write-in Workbook linked to the Student's Book
- this comprehensive Teacher's Guide, with clear suggestions for using the course materials.
- digital materials including slideshows, video clips, additional photographs and interactive activities for use in the classroom.

Approach

The course is designed with student-centred learning at its heart. The students conduct investigations with guidance and support from their teacher. Their investigations respond to questions asked by the teacher or asked by the students themselves. They are practical and activity-based, and include observing, questioning, making and testing predictions, collecting and recording simple data, observing patterns and suggesting explanations. Plenty of opportunity is provided for the students to consolidate and apply what they have learned and to relate what they are doing in science to other curriculum areas and the environment in which they live.

Much of the students' work is conducted as paired work or in small groups, in line with international best practice. Activities are designed to be engaging for students and to support teachers in their on-going assessment of student progress and achievement. Each lesson is planned to support clear learning objectives and outcomes, to provide students and teachers with a good view of the learning. The activities within each unit provide opportunities for oral and written feedback by the teacher, peer teaching and peer assessment within small groups.

Throughout the course, there is a wide variety of learning experiences on offer. The materials are structured so that they do not impose a rigid structure but rather provide a range of options linked to the learning objectives. Teachers are able to select from these to provide an interesting, exciting and appropriate learning experience that is suited to their particular classroom situations.

Differentiation

Differentiation is clearly built into the lesson plans in this Teacher's Guide and levels are indicated against the Student's Book activities. You will see that the practical activities offer three levels of differentiated demand. The square activities are appropriate for the level of nearly all of the students. The circle activities are appropriate for the level of most of the students (this is the level students should be achieving for this stage). The triangle activities are appropriate for some students working towards a higher achievement level. Teachers may find that achievement levels vary for different content strands and interest levels. So students who are working at the circle level in Biology, for example, may find Chemistry topics more interesting and/or easier, so they may work at a different level for some of the time.

Teacher's Guide

This Teacher's Guide is also available to download in editable format at collins.co.uk/internationalresources. Each double-page spread covers one unit in the Student's Book. Each unit has a clear structure identified by the *Introduction–Teaching and learning activities–Consolidate and review* sequence.

Resources the teacher will require for this unit.

Safety notes and any other useful notes for the teacher appear here.

The main **learning objectives** for this unit.

Thinking and working scientifically from the Cambridge Primary Science curriculum framework covered in the unit are provided as a useful reference for the teacher.

Classroom equipment the teacher will require for this unit.

Key words are repeated from the Student's Book page for the teacher to reinforce during the unit.

Scientific background – a brief summary of the science background that the teacher may find useful for this unit. Scientific background ensures teachers understand the topic and can identify and correct misconceptions as they arise. Teachers are also prompted to elicit and correct misconceptions at appropriate points within teaching sequences.

Physics • Topic 4 Forces and sound

4.8 Loud and quiet sounds

Student's Book pages 70–71

Physics learning objectives

- *Light and sound:* 1Ps.02 Explore that as sound travels from a source it becomes quieter.

Thinking and working scientifically

- *Scientific enquiry: purpose and planning:* 1TWSp.02 Make predictions about what they think will happen.
- *Carrying out scientific enquiry:* 1TWSc.02 Use given equipment appropriately; 1TWSc.04 Follow instructions safely when doing practical work; 1TWSc.05 Collect and record observations and/or measurements by annotating images and completing simple tables.
- *Scientific enquiry: analysis, evaluation and conclusions:* 1TWSa.01 Describe what happened during an enquiry and if it matched their predictions.

Resources

- Workbook pages 62, 63, 64 and 65
- PCM P12: Making a shaker

Classroom equipment

- a selection of different musical instruments
- stopwatch
- materials to make shakers: plastic bottles with lids, funnels, lentils, pasta, small stones, rice, etc., to put into the bottles

Key words

loud • quiet

Be aware of any students who may have a hearing impairment, and adapt the lesson accordingly. If the students use the internet, ensure they do so safely and under adult supervision. Supervise the students when they use scissors.

Scientific background

Sound is the movement of energy through a gas, liquid or solid in longitudinal waves. Sound is produced when energy causes an object to vibrate. The more energy is put in, the louder the sound will be, for example hitting a drum, blowing a whistle.

Our voices act like a musical instrument. Air from our lungs is passed over the vocal cords, which vibrate to make a sound. We need to have enough air in our lungs to sing loudly, or at all. Having more breath allows us to sing for longer or more loudly. Singing loudly pushes more air over the vocal chords, but takes more energy.

At this stage, the students will learn that we hear sounds with our sense of hearing and our ears are the sense organ for this. Some students may start to ask questions about how sound travels to the ear; this should be explained in simple terms as the science will be covered in much greater detail in Stage 5 of this course.

Introduction

- Remind the students of the musical instrument investigation they did in the previous lesson. Ask them to describe the sounds that the shakers made. Lead them to use the key words *loud* and *quiet*. Suggest some other words such as *noisy*, and write these on the board.
- Let the students explore a selection of musical instruments. Ask them if they can make quiet sounds and loud sounds using the same instrument. Ask: *What did you do to make the loud sounds?* Answers will vary depending on the instruments, but you should elicit that they have to be struck, blown or plucked harder (with more force). Repeat, asking: *What did you do to make the quiet sounds?*
- Ask: *Is there another way to make the sounds louder or quieter?* Suggest that you could move closer to or further away from the source of the sound. Let the students tell the class about their experiences of hearing loud noises but from far away, for example an airplane, a music concert. Some students may struggle with this concept and say that the sound is quiet. Accept this as an answer at this stage. (Students will learn more about sound in Stage 5 of this course.) Ask the students to look at Student's Book pages 70–71 and answer the questions as a class.

70

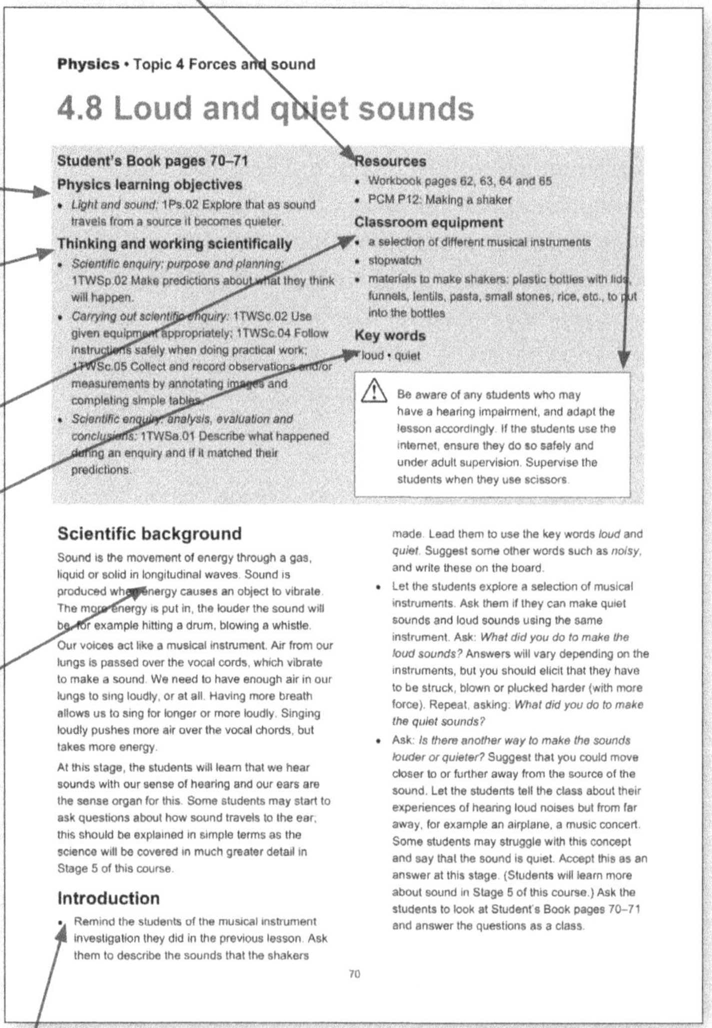

Introduction – this is the introductory part of the unit where ideas are beginning to be explored and students reflect on prior learning and share objectives.

Teaching and learning activities – this leads into the main lesson.

Graded activities – these are differentiated to suit three different levels. They will often involve an investigation and practical element.

The **Consolidate and review** section is used to reinforce the students' learning during the lesson.

Activities with a focus on **language skills and development** are highlighted with a speech bubble icon. These can serve as an ongoing record of important science terms and how they are used. The activities are useful language development exercises, which are important in the international context.

Differentiation – this section discusses the differentiated learning outcomes and provides the teacher with an idea of the likely behaviours of students working at different levels, referencing the square, circle and triangle icons which are used across the course.

Links to the **Collins Big Cat** reading scheme are provided to relate science activities to the English that the students are learning.

At the end of each Topic the answers to the Student's Book questions and Topic quizzes are given in full.

At the back of this Teacher's Guide are the Photocopy Masters (PCMs) and Topic quizzes. These can be handed out to the students as necessary.

Student's Book

Each double page spread covers one unit. Each page has photographs or graphics to provide a stimulus for discussions and questions. There are four double page spreads which cover the key Thinking and Working Scientifically skills. There are also Science in Context case study spreads throughout the book focusing on the history, uses, users or environmental impact of a particular aspect of the topic.

Key words – these are the words that the students will learn and use for this unit.

Questions – these can be used as whole-class discussion points and also to enable the teacher to assess how well individual students understand the unit.

Student's Book

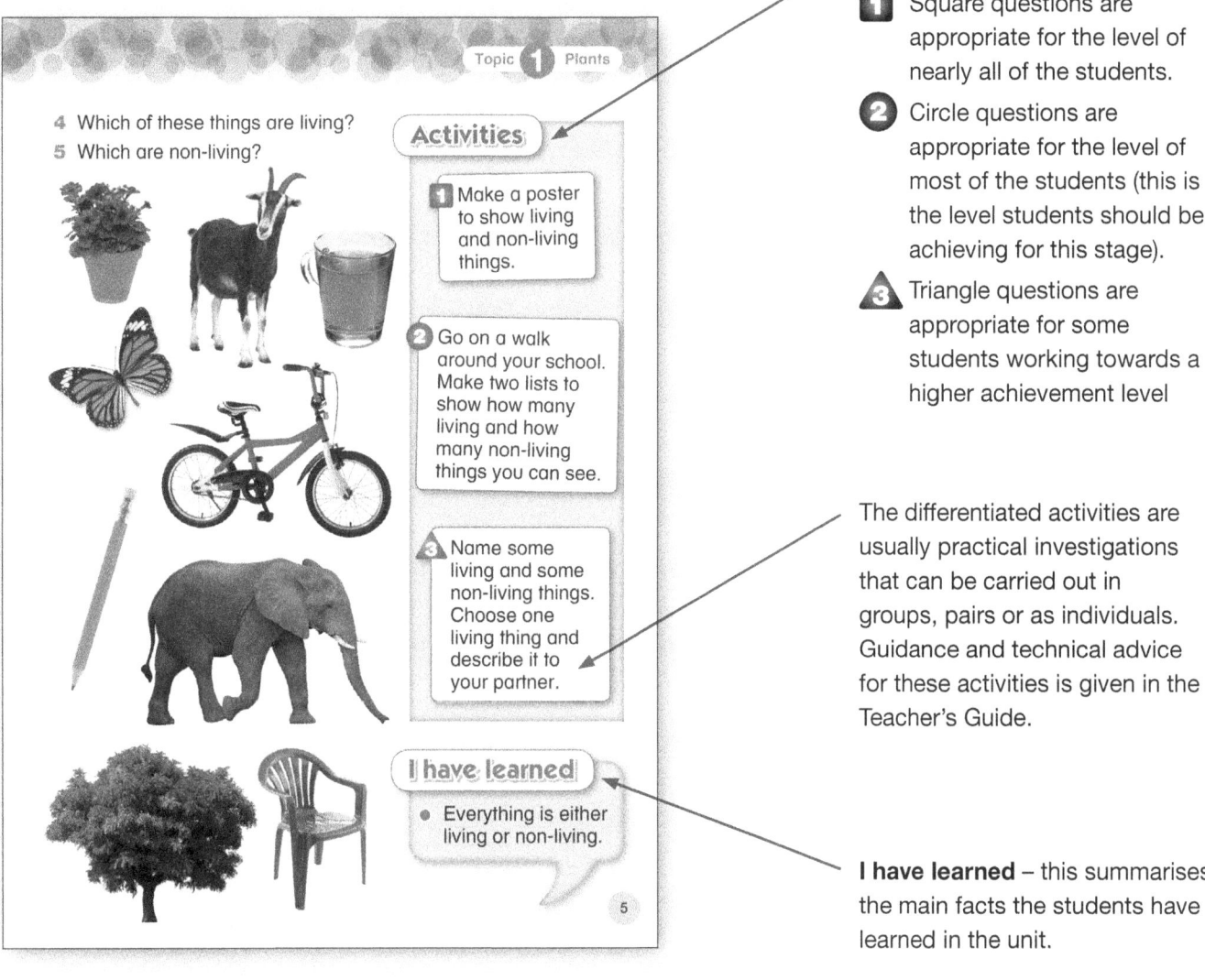

Activities

1. Square questions are appropriate for the level of nearly all of the students.
2. Circle questions are appropriate for the level of most of the students (this is the level students should be achieving for this stage).
3. Triangle questions are appropriate for some students working towards a higher achievement level

The differentiated activities are usually practical investigations that can be carried out in groups, pairs or as individuals. Guidance and technical advice for these activities is given in the Teacher's Guide.

I have learned – this summarises the main facts the students have learned in the unit.

At the back of the Student's Book is a comprehensive **Glossary** of all the Key words that are used during the lessons.

Workbook

The Workbook is for students to record observations, investigation results and key learning during the lesson. It has structured spaces for the students to record work and guidance on what to do. It gives the teacher an opportunity to give the student written feedback and becomes part of each student's work portfolio. There are also opportunities for students to use scientific vocabulary they have learned and develop their understanding of key terms.

The Workbook is intended to be used alongside the Student's Book and Teacher's Guide and therefore it does not cover all aspects of the Cambridge Primary Science curriculum framework.

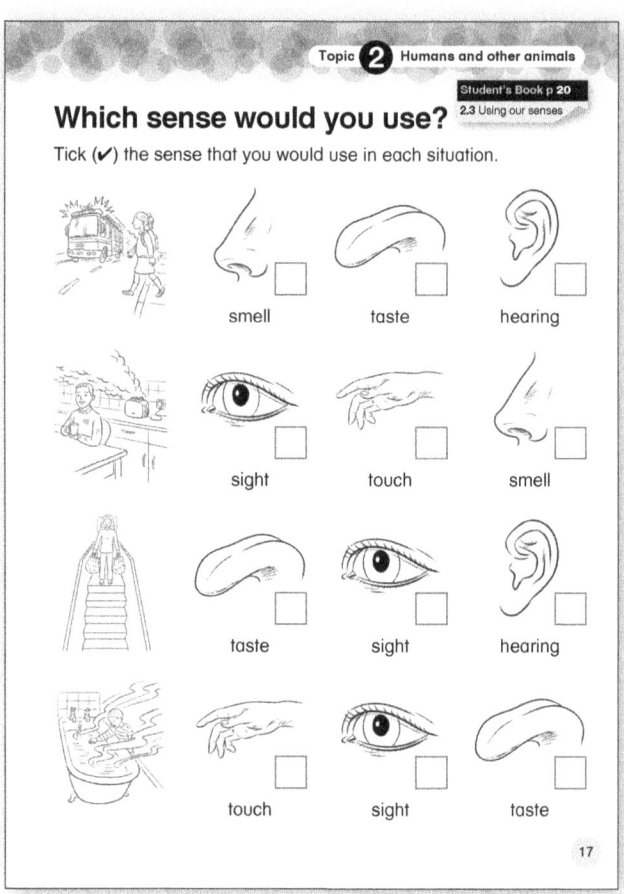

Digital resources

A package of digital resources is available from collins.co.uk to support teachers with learning and assessment. The lesson plans in this Teacher's Guide give references in the *Resources* section and in the body of text to the relevant video clips, slideshows and interactive 'drag and drop' activities.

Interactive 'drag and drop' activities

Slideshows and video clips

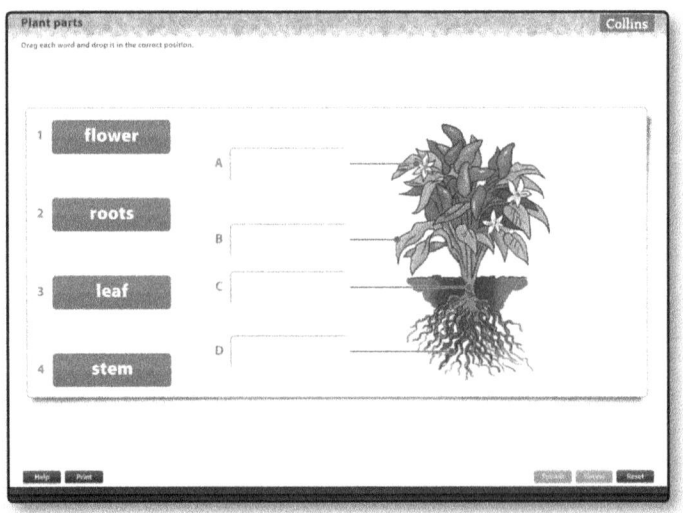

Cambridge Global Perspectives™

Cambridge Global Perspectives is a unique programme that helps learners develop outstanding transferable skills, including critical thinking, research and collaboration. The programme is available for learners aged 5–19, from Cambridge Primary through to Cambridge Advanced. For Cambridge Primary and Lower Secondary learners, the programme is made up of a series of Challenges covering a wide range of topics, using a personal, local and global perspective. The programme is available to Cambridge schools but participation in the programme is voluntary. However, whether or not your school is involved with the programme, the six skills it focuses on are relevant to all students in the modern world. These skills are: research, analysis, evaluation, reflection, collaboration and communication.

More information about the Cambridge Global Perspectives programme can be found on the Cambridge Assessment International Education website:

www.cambridgeinternational.org/programmes-and-qualifications/cambridge-global-perspectives.

Collins supports Cambridge Global Perspectives by including activities, tasks and projects in our Cambridge Primary and Lower Secondary courses which develop and apply these skills. Note that the content of the activities is not intended to correlate with the specific topics in the Cambridge Global Perspectives Challenges; rather, they encourage practice and development of the Cambridge Global Perspectives to support teachers in integrating and embedding them into students' learning across all school subjects.

Activities in this book that link to the Cambridge Global Perspectives are listed at the back of this book on page 167.

Assessment in Cambridge Primary science

In the Cambridge Primary Science curriculum framework, formative assessment is a continuous, planned process that involves collecting information about student progress and learning in order to provide constructive feedback to students and parents, but also to inform planning and the next teaching steps.

The Cambridge Primary Science curriculum framework makes it clear what the students are expected to learn and achieve at each level. Our task as teachers is to make sure that we assess whether (or not) the students have achieved the stated goals using clearly focused, varied, reliable and flexible methods of assessment.

In the *Collins International Primary Science* course, assessment is continuous and in-built. It applies the principles of international best practice and ensures that assessment:

- is ongoing and regular
- supports individual achievement and allows for the students to reflect on their learning and set targets for themselves
- provides feedback and encouragement to the students
- allows for the integration of assessment into activities and classroom teaching by combining different assessment methods, including observations, questioning, self-assessment and informal tasks
- uses strategies that cater for the variety of student needs in the classroom (language, physical, emotional and cultural), and acknowledges that the students do not all need to be assessed at the same time or in the same way
- allows for assessment including controlled activities and tasks. Students should not be tested at all at Stages 1 and 2, and up to Stage 5, formative assessment is preferable to summative.

Assessing scientific enquiry skills

The development of scientific enquiry skills needs to be monitored. You need to check that the students acquire the basic skills as you teach and make sure that they are able to apply them in more complex activities and situations later on.

You can do this by identifying the assessment opportunities in different enquiry-based tasks and by asking appropriate informal assessment questions as the students work through and complete the tasks.

For example, the students may be involved in an activity where they are expected to plan and carry out a fair test investigating cars and ramps (Plan fair test investigations, identifying the independent, dependent and control variables). As the students work through the activity you have the opportunity to check whether they are able to identify:

- one thing that will change
- what things they will measure and record
- what things will be kept the same.

Once they have completed the task, you can ask some informal assessment questions, such as:

- Is a test the only way to do a scientific investigation? (*No, there are other methods of collecting and recording information, including using secondary sources.*)
- Is every test a fair test?
- Are there special things we need to do to make sure a test is fair?
- What should we do before we can carry out a fair test properly? (*Develop and write up a plan.*)
- Is a fair test in science the same as a written science or maths test at school? How is it different?

Registered Cambridge International Schools benefit from high-quality programmes, assessments and a wide range of support so that teachers can effectively deliver Cambridge Primary.

Visit www.cambridgeinternational.org/primary to find out more.

Assessment in primary science

Consolidation and reinforcement

The *Collins International Primary Science* course offers opportunities for consolidation and reinforcement in the form of Topic quizzes that teachers can use to assess learning. These Topic quiz sheets include questions posed in different ways; for example, questions where the students fill in answers or draw diagrams and true or false questions among others.

Below are some examples of the types of questions provided on the Topic quiz sheets. The Topic quiz sheets can be found at the back of this Teacher's Guide.

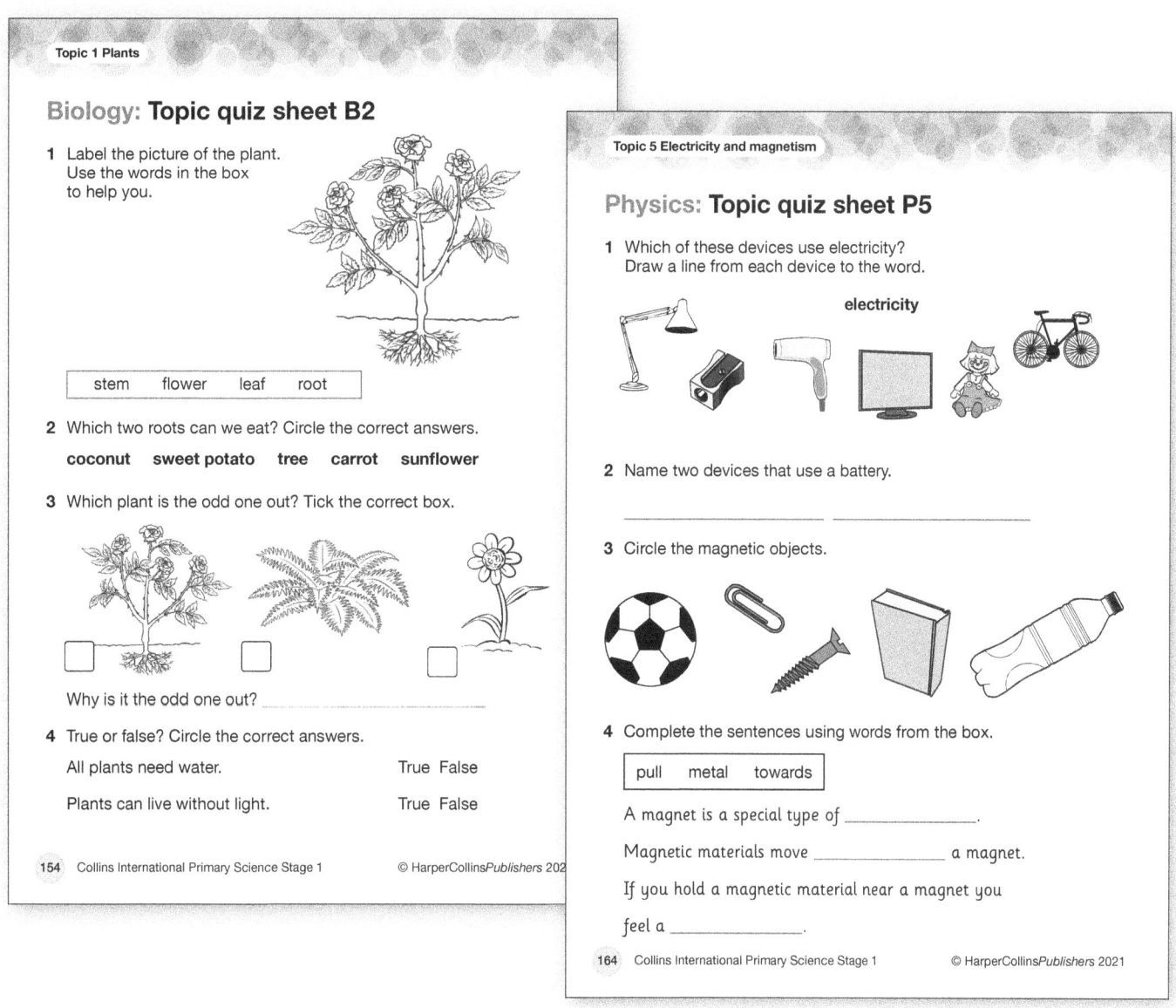

Learning objectives matching grid

Stage 1 Learning objectives	Topic	Unit	Teacher's Guide pages
Thinking and Working Scientifically			
Scientific enquiry: purpose and planning			
1TWSp.01 Ask questions about the world around us and talk about how to find answers.	1	1	2–3
	3	3	36–37
	3	4	38–39
	3	8	46–47
	4	3	60–61
	6	2	86–87
	6	5	93–93
1TWSp.02 Make predictions about what they think will happen.	1	1	2–3
	1	6	12–13
	2	7	28–29
	3	2	34–35
	3	5	40–41
	3	7	44–45
	3	8	46–47
	3	9	48–49
	3	10	50–51
	4	4	62–63
	4	5	64–65
	4	7	68–69
	4	8	70–71
	4	9	72–73
	5	2	78–79

Carrying out scientific enquiry			
1TWSc.01 Sort and group objects, materials and living things based on observations of the similarities and differences between them.	1	3	6–7
	1	4	8–9
	2	5	24–25
	2	6	26–27
	3	1	32–33
	3	2	34–35
	3	6	42–43
	3	7	44–45
	4	2	58–59
	4	3	60–61
	4	7	68–69
	5	1	76–77
	6	4	90–91
1TWSc.02 Use given equipment appropriately.	3	2	34–35
	4	1	56–57
	4	4	62–63
	4	5	64–65
	4	8	70–71
1TWSc.03 Take measurements in non-standard units.	2	7	28–29
	3	2	34–35
	4	1	56–57
	4	4	62–63
	4	9	72–73

These learning objectives are reproduced from the Cambridge Primary Science curriculum framework (0097) from 2020. This Cambridge International copyright material is reproduced under licence and remains the intellectual property of Cambridge Assessment International Education.

Learning objectives matching grid

1TWSc.04 Follow instructions safely when doing practical work.	2	2	18–19
	2	7	28–29
	3	2	34–35
	3	7	44–45
	3	8	46–47
	3	9	48–49
	3	10	50–51
	4	1	56–57
	4	4	62–63
	4	5	64–65
	4	8	70–71
	4	9	72–73
	5	2	78–79
1TWSc.05 Collect and record observations and/or measurements by annotating images and completing simple tables.	1	5	10–11
	1	6	12–13
	2	1	16–17
	2	2	18–19
	2	3	20–21
	3	2	34–35
	3	5	40–41
	3	7	44–45
	3	8	46–47
	3	9	48–49
	3	10	50–51
	4	1	56–57
	4	2	58–59
	4	4	62–63
	4	5	64–65
	4	6	66–67
	4	7	68–69
	4	8	70–71
	4	9	72–73
	5	2	78–79
	6	4	90–91
	6	5	92–93

Scientific enquiry: analysis, evaluation and conclusions			
1TWSa.01 Describe what happened during an enquiry and if it matched their predictions.	1	1	2–3
	1	6	12–13
	2	7	28–29
	3	2	34–35
	3	7	44–45
	3	8	46–47
	3	9	48–49
	3	10	50–51
	4	4	62–63
	4	5	64–65
	4	7	68–69
	4	8	70–71
	4	9	72–73
	5	2	78–79
	6	1	84–85

Con* = Consolidation

These learning objectives are reproduced from the Cambridge Primary Science curriculum framework (0097) from 2020. This Cambridge International copyright material is reproduced under licence and remains the intellectual property of Cambridge Assessment International Education.

Biology	Topic	Unit	Teacher's Guide pages
Structure and function			
1Bs.01 Recognise and name the major parts of familiar flowering plants (limited to roots, leaves, stems and flowers).	1	5	10–11
	1	Con*	14
1Bs.02 Identify the senses (limited to sight, hearing, taste, smell and touch) and what they detect, linking each to the correct body part.	2	2	18–19
	2	3	20–21
	2	4	22–23
	2	Con*	30
1Bs.03 Recognise and name the major external parts of the human body.	2	1	16–17
	2	Con*	30
Life processes			
1Bp.01 Identify living things and things that have never been alive.	1	2	4–5
	1	3	6–7
	1	4	8–9
	1	Con*	14
1Bp.02 Know that animals, including humans, need air, water and suitable food to survive.	2	5	24–25
	2	Con*	30
1Bp.03 Know that plants need light and water to survive.	1	6	12–13
	1	Con*	14
1Bp.04 Describe how humans are similar to and different from each other.	2	6	26–27
	2	7	28–29
	2	Con*	30

Chemistry	Topic	Unit	Teacher's Guide pages
Materials and their structure			
1Cm.01 Identify, name, describe, sort and group common materials, including wood, plastic, metal, glass, rock, paper, and fabric.	3	4	38–39
	3	5	40–41
	3	6	42–43
	3	7	44–45
	3	11	52–53
	3	Con*	54
1Cm.02 Understand the difference between an object and a material.	3	2	34–35
	3	4	38–39
	3	11	52–53
	3	Con*	54
Properties of materials			
1Cp.01 Understand that all materials have a variety of properties.	3	2	34–35
	3	3	36–37
	3	5	40–41
	3	11	52–53
	3	Con*	54
1Cp.02 Describe common materials in terms of their properties.	3	4	38–39
	3	5	40–41
	3	11	52–53
	3	Con*	54
Changes to materials			
1Cc.01 Describe how materials can be changed by physical action, e.g. stretching, compressing, bending and twisting.	3	8	46–47
	3	9	48–49
	3	10	50–51
	3	Con*	54

Con* = Consolidation

These learning objectives are reproduced from the Cambridge Primary Science curriculum framework (0097) from 2020. This Cambridge International copyright material is reproduced under licence and remains the intellectual property of Cambridge Assessment International Education.

Learning objectives matching grid

Physics	Topic	Unit	Teacher's Guide pages
Forces and energy			
1Pf.01 Explore, talk about and describe the movement of familiar objects.	4	2	58–59
	4	Con*	74
1Pf.02 Describe pushes and pulls as forces.	4	3	60–61
	4	4	62–63
	4	Con*	74
1Pf.03 Explore that some objects float and some sink.	4	5	64–65
	4	Con*	74
Light and sound			
1Ps.01 Identify different sources of sound.	4	6	66–67
	4	7	68–69
	4	Con*	74
1Pfs.02 Explore that as sound travels from a source it becomes quieter.	4	8	70–71
	4	9	72–73
	4	Con*	74
Electricity and magnetism			
1Pe.01 Identify things that require electricity to work.	5	1	76–77
	5	3	80–81
	5	Con*	82
1Pe.02 Explore, talk about and describe what happens when magnets approach and touch different materials.	5	2	78–79
	5	Con*	82

Earth and Space	Topic	Unit	Teacher's Guide pages
Planet Earth			
1ESp.01 Know that Earth is mostly covered in water.	6	2	86–87
	6	3	88–89
	6	Con*	94
1ESp.02 Describe land as being made of rock and soil.	6	4	90–91
	6	Con*	94
Earth in space			
1ESs.01 Know that Earth is the planet on which we live.	6	2	86–87
	6	3	88–89
	6	Con*	94
1ESs.02 Describe the Sun as a source of heat and light and as one of many stars.	6	5	92–93
	6	Con*	94
Science in context			
1SIC.01 Talk about how some of the scientific knowledge and thinking now was different in the past.	5	3	80–81
1SIC.02 Talk about how science explains how objects they use, or know about, work.	3	11	52–53
1SIC.03 Know that everyone uses science and identify people who use science professionally.	2	4	22–23
1SIC.04 Talk about how science helps us understand our effect on the world around us.	6	3	88–89

Con* = Consolidation

These learning objectives are reproduced from the Cambridge Primary Science curriculum framework (0097) from 2020. This Cambridge International copyright material is reproduced under licence and remains the intellectual property of Cambridge Assessment International Education.

Lesson plans

Biology	2
Chemistry	32
Physics	56
Earth and Space	84

Biology • Topic 1 Plants

Thinking and working scientifically

1.1 All about Science

Student's Book pages 2–3

Thinking and working scientifically
- *Scientific enquiry: purpose and planning:* 1TWSp.01 Ask questions about the world around us and talk about how to find answers; 1TWSp.02 Make predictions about what they think will happen.
- *Scientific enquiry: analysis, evaluation and conclusions:* 1TWSa.01 Describe what happened during an enquiry and if it matched their predictions.

Resources
- Workbook pages 1 and 2

Classroom equipment
- two potted plants

Note: A few days before the lesson, set up two identical plants for the investigation. Give both plants the same amount of light but water only one plant. Label them 'Water investigation'.

Key words
- investigation • prediction

Skills and connections

Exploring science often begins with a question that can be answered through investigative work. Students need to learn how to choose an appropriate question and then decide how best to answer it. They should ask themselves 'What do I want to find out?' and 'How can I find this out?' For example, in this unit, students will think about what plants need to survive and should be encouraged to select questions such as 'Do plants need water?' and 'What happens if plants do not get water?' Students can then be directed towards ideas for investigations that will enable them to answer such questions.

Making predictions is a key skill that students will be expected to apply across the curriculum as they progress through their education. At this stage, it is most often applied in Science (*What will happen if we don't water the plant?*) and in English (*What do you think will happen next in the story?*). Students will make predictions in their everyday lives too (*What might happen if I don't...?*) without necessarily being aware of this.

In its simplest form, making a predication is answering the questions: *What do you think will happen?* and *Why?* In order to be able to do this, other skills need to be utilised, such as applying existing knowledge of the world and understanding what is happening in a given situation. These skills are combined to arrive at an educated guess (a prediction) of what the outcome will be. As well as being able to predict what might happen, students need to be able to give sound reasons for their prediction, giving as much detail as possible and using science words as far as they can. At this stage, they can do this in English or their native language if they prefer. The key thing is that they practise using the skills required to make sensible predictions. You can provide them with additional vocabulary as necessary.

The skills of asking and answering questions, and making predictions, will be revisited throughout the course in a variety of contexts. For example, in Unit 1.6 students will predict how light affects the growth of plants and in Topic 3 they will ask questions and make predictions about the properties of materials. They will also be expected to use these skills as they do different investigations of their own.

Introduction

- Tell students that they are going to play a prediction game. Explain that you are going to act out some everyday activities and give the class some clues. Some examples could be picking up a book and sitting down on your chair (you are going to read), wiping your forehead to indicate that you are hot and picking up a water bottle (you are going to have a drink), picking up a pen and standing near the board (you are going to write on the board), picking up a ball and walking towards the door (you are going outside to play), etc.
- Students need to guess what you are going to do next and give reasons. Remind them to use the sentence structure 'I think…because…' You could work through one as an example first. Help

Biology • Topic 1 Plants

the class by explaining the steps involved in making a prediction. Remind them to look at the clues and use what they know about the world.

- After you have done two or three scenarios, students can take turns working in groups to do some more. Circulate to check that students are using the correct language and giving reasons for their predictions.
- Explain that it does not matter whether a prediction is correct or not. The important thing is that students try to apply their knowledge of the world and give sensible ideas for what they think the outcome might be. Tell them that in Science, we often start with a question, make a prediction and then do an investigation to find out whether the prediction was correct or not.

Teaching and learning activities

- Turn to Student's Book page 2. Work through the questions as a class. Students' answers will show their prior knowledge before you start this topic on plants. Some students may be able to answer using the names of the plant parts (roots, leaves, stems, flowers) and give ideas for what plants need to survive. Their accuracy, or ability to do this in English, is not critical. The priority is that they begin to think about plants and what plants need.
- Tell students to look at the picture on Student's Book page 3. Ask the class questions to guide them to the correct answers: *What do you notice about the labels on the two plant pots? What do you think the boy in the white shirt's job is? What are the children doing?* (watering one of the plants) *Will both plants get water?* (no) *What question do you think the children are investigating?* (Do plants need water?) *What might happen to the plant without water?* (It might die.)
- Discuss ideas as a class. Remind students to give reasons for their predictions and to use science words as much as they can.

Graded activities

1 Let students complete the activity on Workbook page 1. Some may need help to explain their predictions. Circulate to ensure that students are using the structure 'I think…because…'

2 Students can work in pairs to discuss what they think the two plants will look like. This is a good opportunity for them to use their prior knowledge and vocabulary for plant parts and what plants need to survive. Ask different pairs to share their ideas with the class. Make a class list on the board.

3 Explain to students that you have been growing two different plants: one you have been watering and the other you have not watered. Before you show the plants to the students, ask: *What do you think the plant will look like if it has had no water?* Let them discuss this in groups. They should draw pictures to show their ideas and make predictions on Workbook page 2. Show the two plants to the class. Ask the students to observe the differences and to discuss them in groups. To finish, ask: *Do plants need water?*

Consolidate and review

- Ask students to work in pairs to describe scenarios to each other and then say what they think will happen next. These do not need to be science contexts. The aim is to practice 'I think… because…', giving detail to explain their reasons.

Differentiation

■ All of the students should be able to predict what will happen next in each situation. Encourage students to explain their reasons in detail. If they struggle to do this, offer support.

● Most of the students should be able to describe some differences between the two plants and begin to understand that lack of water affects the growth of a plant. Some students may find this activity easier than others. Circulate, asking questions to guide them to the correct answers.

▲ Some of the students should be able to work collaboratively, asking and answering questions to clarify their thinking. Some should be able to make predictions but may struggle to show this pictorially. Tell students that they are going to learn more about what plants need to survive later in this unit.

Biology • Topic 1 Plants

1.2 Is it alive?

Student's Book pages 4–5

Biology learning objective
- *Life processes:* 1Bp.01 Identify living things and things that have never been alive.

Resources
- Workbook page 3
- Slideshow B1: Is it living or non-living?
- PCM B1: Living or non-living?

Classroom equipment
- large sheets of paper
- coloured pens or pencils
- scissors
- glue
- selection of old magazines
- a doll

Key words
• living • non-living

 Supervise the students when they are cutting with scissors and working with glue. If you take the students on a walk around the school grounds, ensure they are safe and that they stay together. They should not put anything they find into their mouths.

Scientific background

Everything in the world can be sorted into *living* or *non-living*. In Unit 1.3, the students will learn that if something is living it is either a *plant* or an *animal*. In Unit 1.4, the students will learn that non-living things can be sorted into those that *once lived* and those that *have never lived*. For this lesson, they do not need to know any of these distinctions, but should be able to identify whether something is living or non-living.

Introduction

- Use the topic opener photograph on Student's Book page 1 as a talking point. Ask the students to describe what they can see. Let them briefly discuss in groups what they know about plants and flowers. Tell them that they are going to learn all about plants in this topic.
- Ask the class to look at the picture on Student's Book page 4. Ask: *What can you see in the picture? Can you suggest some groups you could sort the things in the picture into?* Take feedback but guide the students to select *living* and *non-living* as the criteria.
- Introduce the terms *living* and *non-living* as key words. Ask the students to repeat the words after you. Invite them to give examples of living and non-living things. This will help you assess their prior knowledge and ability to differentiate between living and non-living. Finally, repeat the key words again.

Teaching and learning activities

- Make sure the students understand the questions on Student's Book pages 4–5. Remind them of the two groups: *living* and *non-living*. Ask them to identify all the living and non-living things in the pictures. Then let them discuss their answers in groups. Some students may begin to talk about the similarities and differences they notice between living and non-living things. You can encourage this to develop their interest in the topic, but do not correct any misconceptions at this stage. Check the answers as a class.
- Ask the students to cut out pictures of living and non-living things from magazines. Allow them time to freely discuss the pictures with a partner, saying which pictures they chose and why. The pictures can be kept for using on the posters the students will make in the first graded activity.

Graded activities

1 Give each pair of students a large sheet of paper, scissors, glue, coloured pens or pencils and a copy of PCM B1, which shows a variety of living and non-living things. Ask the students to sort the pictures into two groups: *living* and

Biology • Topic 1 Plants

non-living. Once the sorting is done, ask them to make a poster by sticking the pictures onto the large sheet of paper. Encourage the students to colour in the pictures, as this develops fine motor skills and gives practice at holding a pencil. They should write *living* or *non-living* next to each picture. Ask: *Which things are living and which are non-living?* Stick the finished posters on the wall and discuss them with the students.

❷ If possible, take the students on a walk around the school grounds. Say: *Count how many living things you can see. Count how many non-living things you can see.* Encourage them to make lists and to classify the objects as being either living or non-living. Ask: *What made you decide if a thing was living or non-living?* (Before the lesson you could 'hide' some living and non-living objects around the school grounds for the students to find.)

❸ Show the class Slideshow B1 of living and non-living things and ask: *Which are the living things? Which are the non-living things?* Let them discuss their answers in pairs. Take feedback and invite the class to name each thing in the slideshow, either in English or in their own language. Then ask the students to choose one living thing that they know the name of and to describe it to a partner. Some students may prefer to make their choice from those shown on the slideshow. Students working at a higher level can choose any appropriate living thing.

Consolidate and review

- Use Workbook page 3 to consolidate the teaching and to check that the students can distinguish between living and non-living things.
- Show the students a doll and ask them to suggest, to a partner, the ways in which they are similar to the doll and the ways in which they are different. Encourage them to conclude that they are living and the doll is non-living. Some students may begin to discuss the characteristics of living things (they will learn more about these in Stage 3 of this course).

Differentiation

■ All of the students should be able to sort the pictures into living and non-living things. Some will be able to do this more quickly than others, but they can move onto the colouring and sticking. Some students may like to add the pictures that they cut from magazines to their poster and should be able to label these correctly. If some students need support with the labelling, write *living* and *non-living* for them to copy.

● Most of the students should be able to identify a selection of living and non-living things during the walk with a little help. If not, demonstrate by showing them a few examples. Most students should be able to record what they see and to organise the things into two lists: one for living and one for non-living. This is useful practice in a method for recording data.

▲ Some of the students should be able to identify the specific examples of living and non-living things and give a detailed description of a living thing of their choice to a partner. Encourage them to describe as many features as possible, for example size, colour and shape. You could turn this into a guessing game by asking students to hide the identity of the living thing they have chosen; partners should try and guess what it is, based on the description.

BIG CAT 🐾

Students who have read *Collins Big Cat: The Oak Tree* may relate the learning in this lesson to the many different types of organism, including plants and animals, described in this book. It provides a lively introduction to the topic of living things.

Biology • Topic 1 Plants

1.3 Plants and animals are living things

Student's Book pages 6–7

Biology learning objectives
- *Life processes:* 1Bp.01 Identify living things and things that have never been alive.

Thinking and working scientifically
- *Carrying out scientific enquiry:* 1TWSc.01 Sort and group objects, materials and living things based on observations of the similarities and differences between them.

Resources
- Workbook page 4
- Slideshow B2: Living and non-living
- PCM B2: Plants or animals? (1 and 2)

Classroom equipment
- large sheets of paper
- coloured pens or pencils
- scissors
- pictures or specimens of plants and animals
- large cards

Key words
- plant • animal • alive

 Supervise the students when they use scissors.

Scientific background

Plants and animals are living things. If something is alive it can be seen at this stage as either a plant or an animal. All living things can be sorted into groups according to their features. The two main groups are plants and animals. The sorting of living things into either plant or animal is a very simplistic form of grouping. Students will learn in later stages that these two groups can be divided into smaller groups. For example, animals can be divided into vertebrates and invertebrates, and then further divided into groups such as mammals, birds and fish. They do not need to know these detailed classifications yet, but should be able to classify living things into the two basic groups: plants and animals.

Introduction

- Write the key words on the board: *plant*, *animal* and *alive*. Explain that we say a living thing is *alive*. Ask the students to repeat each key word after you. The students should be familiar with the concept of plants and animals in a general sense and be able to name examples of each. Invite individual students to name a plant or animal, in their own language or in English, and write their responses on the board in two separate lists. Introduce new words as necessary.

- Remind the students that we can classify everything around us into *living* and *non-living*. Show the students Slideshow B2 about living and non-living things. Say: *Can you suggest how you would sort the living things into groups?* Take feedback but guide the students to select *plants* and *animals* as the criteria. Explain that plants and animals are living things: they are alive. Ask the students to say whether each living thing is a plant or an animal. Once they have identified the living things, ask if they know the names of any of the plants and animals, and add these names to your class lists on the board.

- Ask the class to look at the picture on Student's Book pages 6–7. Ask: *Which things in the picture are living? Which are non-living?* Let them discuss this in their groups. Ask: *What made you decide if a thing was living or non-living?* Take feedback and write their responses on the board. Say: *Count how many butterflies you can see in the picture. Count how many birds you can see.* Counting the animals out loud will help to reinforce numeracy skills.

Teaching and learning activities

- Have available lots of pictures of animals and plants or, even better, specimens of the real things. In groups of three, the students should sort them according to their own criteria. As they

Biology • Topic 1 Plants

do so, circulate and check that they are sorting correctly. Ask: *Can you explain how you sorted the plants and animals into groups?* Different groups of students should be challenged to explain why they have grouped things in a particular way.

- Make sure students understand the questions on Student's Book pages 6–7. Discuss the answers as a class. Ask the students how many different types of plants are in the picture and to compare their similarities and differences. Circulate, asking questions to draw out some simple similarities and differences. Ask: *What colour are the plants? Where do the plants live? Are the plants all the same size?*

- Ask: *Can you explain how we know if something is a plant or an animal?* In pairs, ask the students to draw an imaginary animal and to list all the things that it can do. Do not correct any misconceptions at this stage.

- Repeat the exercise with an imaginary plant. The students should be encouraged to think about what colour, shape and size the plant should be. They can also think about where the plant is growing. At this stage, any reasonable suggestions are fine.

Graded activities

1. Give each pair of students a copy of PCM B2. Ask them to cut out the pictures and to sort them into plants and animals. As they do so, circulate and check that they are sorting the pictures correctly. If the pictures are photocopied onto card, they can be used repeatedly. This activity will give you a clear idea of the students' understanding of the grouping of living things.

2. Remind the students about the walk around the school grounds which they did in the previous lesson. Tell them to look back at their list of living things. Ask: *How many living things did you find?* Ask them to write *plant* or *animal* next to each living thing on their list. Some students may be able to name specific plants and animals.

3. Give the students a piece of paper and ask them to draw a line down the middle and label one side 'plants' and the other 'animals'. Then ask them to draw or write about a plant and an animal that they know. They should describe the features and compare them. The drawings can be fictitious, but the descriptions of the features should be accurate. Ask the students: *Has the plant got legs? Does the animal have more than two eyes?* Discuss with the class why they chose certain features. Explain that when we *compare* things, we look for things that are the same and things that are different. Ask: *In what ways are they the same? In what ways are they different?*

Consolidate and review

- Use Workbook page 4 to consolidate the teaching and to check that the students understand that everything is either living or non-living, and that living things can be grouped into plants or animals.

- Play the 'What am I?' game: you say one clue at a time, with the clues leading to the identity of a familiar plant or animal; the students have to guess what it is. Once the rules are established, the students can play the game on their own.

Differentiation

■ All of the students should be able to sort the familiar living things into plants and animals based on obvious criteria.

● Most of the students should be able to identify and correctly label each of the living things on their list as either a plant or an animal, with little prompting. Some may also know the names of the plants and animals that they found.

▲ Some of the students should be able to start comparing similarities and differences between plants and animals by considering their different features. Students working at a higher level should be able to explain their reasoning in simple terms.

Biology • Topic 1 Plants

1.4 Things that have never been alive

Student's Book pages 8–9

Biology learning objective

- *Life processes:* 1Bp.01 Identify living things and things that have never been alive.

Thinking and working scientifically

- *Carrying out scientific enquiry:* 1TWSc.01 Sort and group objects, materials and living things based on observations of the similarities and differences between them.

Resources

- Workbook pages 5 and 6

Classroom equipment

- scissors
- glue
- selection of old magazines
- large sheets of paper

Key words

- non-living • alive • living

 Supervise the students when they are cutting with scissors and working with glue. If the students explore things around the classroom, make sure they do so safely.

Scientific background

Everything in the world can be sorted into *living* or *non-living*. All living things carry out the *seven life processes*: they move, respire, grow, reproduce (have young), feed, excrete waste and are sensitive to changes in their environment (via their senses). Non-living things do not do these things (for themselves). Students do not need to understand this level of detail at this stage, just enough to be able to distinguish between living and non-living.

As with living things, non-living things can be divided into groups. They can be grouped into those which *once lived* and therefore carried out the seven life processes, such as wood used for furniture, and those which *have never lived*, such as stone. Non-living things that once lived come from *natural* sources, such as cotton and fruit from plants, and wool and leather from animals. Things that have never lived can also be found in nature, but many are *manufactured*, such as metal, glass and plastic. Students do not need to differentiate between natural and manufactured materials at this stage, but they should be able to classify non-living things into those that *once lived* and those that *have never lived*.

Introduction

- Reinforce the students' learning of living and non-living by asking: *What is the same about a chair and a person?* (They both have legs.) Ask: *Does that mean the chair is a living thing? Why?* Let the students discuss their ideas in groups. Take feedback as a class.
- Next, ask the class: *Can you tell me the differences between living and non-living things?* Take feedback. At Stage 1, students' main focus will be on characteristics that are familiar to them, such as feeding and growing. Other characteristics of living things will be taught in more detail in Stages 2 and 3 of this course. Write the responses on the board. Choose an object that is definitely living, such as a horse, and check against the ideas on the board. Then do the same with a non-living object. Address any misconceptions.
- Ask the students to look at the scene in the picture on Student's Book pages 8–9. Ask the class questions and guide them to the correct answers: *Where was the wood in the logs before they were used to build the hut?* (on a tree) *What was the tree doing?* (growing) *Was the wood alive?* (yes) *Are the logs growing now?* (no) *What does this tell us?* (They are non-living but were once alive.)

Teaching and learning activities

- Explain that some things have never been alive and illustrate this point with plenty of examples. Tell students that they can work out whether something has been alive or not by thinking about whether it used to grow and need food and water to stay alive. If the object has never done these things, then it has never been alive.

Biology • Topic 1 Plants

- Make sure the students understand the questions on Student's Book page 9. Then let them discuss their answers in groups. They should identify the non-living things in the picture and then classify these into *once lived* and *has never lived*. Circulate, offering support and guidance as necessary. Check the answers as a class. Ask the students to explain their reasoning as fully as possible.
- Identify any misconceptions that the students may have; for example, some students may think that wood is living. Explain: *Wood was once living, but when it is cut from a tree it is no longer growing, so is no longer living.*

Graded activities

1 Let the students complete the activity on Workbook page 5. They should circle all the non-living things that have never been alive. Some students may need help distinguishing between non-living things that were once alive, such as wood, and things that have never been alive, such as the radio. Circulate, asking questions to guide them as necessary, e.g. ask: *Where did the vegetable come from? Was it alive when it was part of the plant? Is it alive now?* When the students have completed the task, ask them to name the non-living things, either in English or in their own language.

2 Ask the students to walk around the classroom. Tell them: *Look for some things that were once living. Look for some things that have never been alive.* Ask them to make a list of as many things as they can. (Before the lesson you could 'hide' some once living and never lived objects around the classroom for the students to find.) This activity should show how well the students have understood the difference between the two categories of non-living things.

3 Put students in pairs or small groups and give them some old magazines, a sheet of paper, scissors and glue. Ask them to make an odd-one-out poster by cutting and sticking pictures of things that were once alive and things that have never been alive. Explain that they should choose several objects from one group and only one from the other group. The students should also suggest why the object is the odd one out and write this on the back of their poster. Use this activity to assess their reasoning.

When they have completed their posters, let other groups try to identify the odd one out.

Consolidate and review

- Use Workbook page 6 to consolidate the teaching and to check that the students understand the difference between living things and non-living things that have never been alive.
- Ask the students: *Can you explain how we know that we are living? What do we do that a doll does not do? What does a flower do that we do?* Allow them time to discuss their ideas in groups. Take feedback and accept any suitable answers.

Differentiation

■ All of the students should be able to identify the non-living things that have never been alive with little prompting. If not, help the students by reminding them of the basic criteria of living things and asking whether each object ever did those things.

● Most of the students should be able to identify a selection of once alive and never been alive things. Most students should be able to record what they see and organise the data in two lists: once alive and never been alive. This is useful practice in a method for recording data.

▲ Some of the students should be able to independently choose a selection of once alive pictures and one never been alive picture, or vice versa. They should be able to work collaboratively, to think creatively and correctly sort objects into groups, and then explain their reasoning.

Biology • Topic 1 Plants

1.5 Parts of a plant

Student's Book pages 10–11
Biology learning objective
- *Structure and function:* 1Bs.01 Recognise and name the major parts of familiar flowering plants (limited to roots, leaves, stems and flowers).

Thinking and working scientifically
- *Carrying out scientific enquiry:* 1TWSc.05 Collect and record observations and/or measurements by annotating images and completing simple tables.

Resources
- Workbook pages 7 and 8
- Slideshow B3: Leaves and flowers
- PCM B3: Five little leaves
- Slideshow B4: Fruit and vegetables
- Digital resource B1: Plant parts

Classroom equipment
- variety of leaves
- variety of plants or flowers
- glue, colouring pens or pencils, large sheets of paper
- variety of edible roots
- paper and wax crayons

Key words
- root • stem • leaf • flower

 The vegetables must be clean and fresh. The students should not eat any plant parts unless supervised. Some leaves can cause irritation. Some students may have allergies to some types of pollen. Warn students not to rub their eyes or faces and not to put their fingers near their mouths. Make sure they wash their hands after handling the plants. Supervise the students when they are working with glue.

Scientific background

All flowering plants have the same parts. The *leaves* use energy from the Sun, carbon dioxide and water to make food for the plant (glucose) and oxygen. *Roots* have two functions: they anchor the plant securely in the soil and absorb water and other nutrients. Most roots grow underground. The *stem* transports the water and nutrients around the plant, similar to an animal's circulatory system. The stem also holds the plant upright. Not all plants produce *flowers*.

Different plants have different kinds of roots. Some plants have *fibrous roots* and other plants have *tap roots*. The tap root may have smaller branches growing off it. Some tap roots, such as carrots, are edible. At this stage, students should know that different plants have different types of roots but they do not need to know their names.

At Stage 1 the students only need to be able to recognise and name the major parts of a flowering plant. They do not need to know about the roles of the different plant parts (they will learn more about these in Stage 3 of this course).

Introduction

- Ask the students: *What do you know about plants?* Encourage them to take turns to share their opinions and listen to the opinions of the others in the group. Write their ideas on the board.
- Show the students a variety of plants or a bunch of assorted flowers, with different colours, sizes, scents, number of leaves. Ask the students to examine the flowers, investigating their colours, scent, feel, number of leaves. As a class, discuss and compare the plants in turn. Ask the students to comment about the colour, texture and leaves, asking questions to guide them: *Are all the plants the same colour? Do the flowers smell the same? Are all the leaves big?* Draw the plants on the board and write the students' comments alongside each drawing.

Teaching and learning activities

- Ask the class to look at the pictures on Student's Book page 10. Introduce the different parts of a plant as key terms: *stem, root, leaf* and *flower*.

Biology • Topic 1 Plants

Write the words on the board and point out the spelling and pronunciation. Ask the students to point to each part. Make the distinction between the name of the whole plant and its parts, e.g. an orchid flower and an orchid plant. Ask the students to make a list of the similarities and differences between the two plants. Remind the students that when we look to see how things are the same and how they are different, we say we are *comparing* them.

- Show the students Slideshow B3 of different leaves and flowers. Pause the slideshow after each image, to allow the students time to discuss each question in groups. This activity encourages the students to start thinking about the fact that plants have parts that are the same, but that the parts can look very different from one plant to the next.

- Ask the students to look at the pictures on Student's Book page 11. Discuss the plant roots and answer the questions as a whole class. Ensure that the students understand that the carrot and the radish are roots that have grown large. Have some root vegetables and fibrous roots available for them to handle and look at. Ask the students what root vegetables they know about, e.g. sweet potato. Establish that some plants have roots we can eat.

- Talk about plants that we can eat and ask the students to name some. Move on to discuss the fact that not all plants can be eaten and some are poisonous. Explain that we should never eat a plant unless we know that it is safe to eat.

Graded activities

1 Let the students complete the activity on Workbook page 7. Circulate and suggest adjectives if the students need prompting. This activity helps to illustrate that plants have the same parts, but that there are differences between these parts. It also helps the students to practise their observational skills.

2 Let the students complete the activity on Workbook page 8. Circulate, asking questions to guide students, e.g. *What is this plant part called?* This activity reinforces the names of the different parts of a plant.

3 Take the students outside to look at a tree (or have pictures of trees available if this is not possible). Ask: *Can you show me the stem of the tree?* This question will stretch some students and they may identify the branches or leaf stems instead of the trunk. Explain that the 'trunk' of a tree is its stem – it is just a lot bigger than most plant stems. Let the students sketch a tree of their choice and label the parts. They can also make rubbings of the leaves and bark using paper and wax crayons. If the students make a variety of different rubbings, they can use these to quiz each other to see if they can identify which tree it comes from. Talk about the many different types of trees, but make sure that the students understand they are all plants despite the differences in their size and shape.

Consolidate and review

- Make a poster of an imaginary tree using real leaves. Provide the students with a trunk and branches made from paper. Ask the students to collect leaves from the school grounds or home. The leaves can be stuck to the branches to produce the 'class tree'.

- Show the students Slideshow B4, of various named fruits and vegetables. Say the name of each fruit or vegetable, as it is displayed. Ask: *Who can make a sentence describing the fruit/vegetable?*

- Let the students complete Digital resource B1 to consolidate their learning.

- Teach the students the *Five little leaves* action rhyme, from PCM B3, and help them to recognise the rhythm and repeating action.

Differentiation

■ All of the students should be able to suggest a range of differences between two plants of their choice, such as colour and shape. Some students should be able to name less obvious differences, such as smell and texture.

● Most of the students should be able to draw a flowering plant and correctly label the key features: stem, root, leaf, flower.

▲ Some of the students should be able to say that the trunk of a tree is its stem. They will be able to make an accurate drawing of their tree and label the parts correctly without any help.

Biology • Topic 1 Plants

1.6 What do plants need to survive?

Student's Book pages 12–13

Biology learning objective
- *Life processes:* 1Bp.03 Know that plants need light and water to survive.

Thinking and working scientifically
- *Scientific enquiry: purpose and planning:* 1TWSp.02 Make predictions about what they think will happen.
- *Carrying out scientific enquiry:* 1TWSc.05 Collect and record observations and/or measurements by annotating images and completing simple tables.
- *Scientific enquiry: analysis, evaluation and conclusions:* 1TWSa.01 Describe what happened during an enquiry and if it matched their predictions.

Resources
- Workbook pages 9, 10 and 11
- Video B1: Roots growing

Classroom equipment
- large sheets of paper, coloured pens or pencils
- potted plant
- two bean plants
- paper
- various large leaves
- paint

Note: A few days before the lesson, set up two identical bean plants for the investigation. Give both plants the same amount of water but keep one in the light and one in the dark. Label them 'Light investigation'.

Key words
- water • light • grow • soil • air • healthy

Scientific background

The great majority of plants can make their own food, which they need in order to live and grow well. The green parts of a plant make the food by the process of *photosynthesis*. Water and light are essential for photosynthesis and therefore essential for a plant to live and grow. Plants' leaves turn naturally towards the light. As plants grow, they get bigger and develop more leaves. Plants grown without light are often tall, thin and pale. They are not strong plants. Without water, plants droop or wilt. If they go without water for too long, they will die. At Stage 1 the students do not need to know the details of photosynthesis, but they do need to understand that all plants need water and light to survive. (The students will learn that plants get their energy from light in Stage 4 of this course.)

In this lesson, the students will do a class investigation to observe how light affects the growth of plants.

Introduction

- Ask the students: *What do you remember about plants?* Give them a few minutes to discuss this in their groups. Encourage them to take turns to share their opinions and listen to the opinions of the others in the group. Take feedback. Encourage the students to discuss their ideas with the class.

- Review the previous lesson by drawing a picture of a plant on the board. Add labels by eliciting the names of the different parts from the class.

- Tell the class that they are now going to learn more about the things that plants need to live and grow well. Introduce the key terms: *water*, *light*, *grow*, *soil*, *air* and *healthy*. Write the words on the board and point out the spelling and pronunciation.

Teaching and learning activities

- Ask the class to look at the picture of three plants on Student's Book page 12. Ask the students about the differences between the plants, and why they think the plants are different. Allow them to discuss their ideas in groups and then take feedback. Establish that one plant is healthy and the other two are unhealthy.

- Show the class Video B1 of roots growing. Ask: *What happens to the roots as the plants grow?* Most students will realise that the roots also grow. Some may comment that the roots grow and move downwards.

Biology • Topic 1 Plants

- Show the students a potted plant. Take it out of the pot to show its roots. Ask the class to observe the colour of the roots and how the roots have grown. Ask: *What do you think plants need in order to grow?* Allow the students to discuss this in groups. Take feedback and write their ideas on the board.

- Explain that plants need water to survive, just like we do. Ask: *What does your mouth and body feel like when you need water? What happens if you cannot find water?* Say that plants also need light to live and grow. Explain, in very simple terms, that plants make their own food. Stress the importance of light (usually from the Sun) and water so they can do this. Without water and light, plants will die. Encourage the students to give their opinions and to ask and answer questions. Address any misconceptions the students may have about plant food coming from the soil.

Graded activities

1 Explain to the students that they are going to make a poster called 'What plants need to survive'. Ask the students, in their groups, to make posters showing everything they know about plants and what plants need in order to live and grow well. Circulate, asking questions to ensure that the students include all the key features. Display the posters on the wall. Let the students take turns to look at the posters. Ask the students to say what is good about each poster. If appropriate, they can also point out any mistakes and say how the posters could have been even better.

2 Tell students they are going to explore how light affects the growth of a plant. Show them a healthy bean plant (the one you have been keeping in the light). Ask: *Can you describe how light affects the growth of plants?* Explain that you have been giving two identical plants the same amount of water, but keeping one in the light and one in the dark. Ask: *What do you think has happened?* Students should draw the plant you showed them, on Workbook page 9, and work in pairs to predict what has happened to the plant that did not get any light.

3 Show the students the two bean plants. Remind them that you gave both plants the same amount of water but kept one in the light and one in the dark. Ask: *Can you describe how the plants are different? What has changed? Why has this happened?* Ask the students to observe the effects of lack of light and to record their observations on Workbook page 10. Discuss the students' ideas and establish that plants need light to make food and therefore to grow well and be healthy. Ask: *Which is the healthiest plant? Why?* Encourage the students to explain their reasoning as fully as possible.

Consolidate and review

- 💬 Use Workbook page 11 to consolidate the teaching and to check that the students understand the things that plants need in order to live and grow well, in particular the importance of water and light.

- Ask the students to work in pairs to tell each other what plants need to survive.

- Let the students use large leaves and paint to make leaf prints. Discuss with the students the different shapes of the leaves and different patterns that they make.

Differentiation

■ All of the students should be able to work together to share ideas in their groups, asking and answering simple questions and supporting each other, in order to produce a poster about the needs of a plant.

● Most of the students should be able to make and draw their predictions with a little support. Most should be able to work collaboratively with a partner to help clarify their thinking. Some may find making predictions easier than others; offer support as necessary.

▲ Some of the students should be able to compare the two plants and think critically to name the differences. Some should be able to accurately describe the results and explain the outcome in simple terms. Some may struggle to link the lack of light with poor growth. If so, help by reminding the class about what plants need to live and grow well.

Biology • Topic 1 Plants Consolidation

Consolidation

Student's Book page 14

Biology learning objectives
- *Structure and function*: 1Bs.01 Recognise and name the major parts of familiar flowering plants (limited to roots, leaves, stems and flowers).
- *Life processes:* 1Bp.01 Identify living things and things that have never been alive; 1Bp.03 Know that plants need light and water to survive.

Resources
- Workbook page 12
- Topic quiz sheets B1 and B2

Classroom equipment
- coloured pens or pencils

Looking back Topic 1

- Use the summary points on Student's Book page 14 to review the key things that the students have learned in the topic. Ask questions such as: *Can you name some living things in the classroom?* (teacher, students, class pets) *Can you name some non-living things in the classroom?* (tables, chairs, books) *Are there any things in the classroom that are non-living but were once alive?* (wood in furniture) *What can you remember about the parts of plants?* (roots, stem, leaves, flowers) *Can plants live without water?* (no) *What else do plants need to survive?* (light)
- Distribute some coloured pens and ask the students to draw the plant of their choice on Workbook page 12. When they have completed their drawings, the students should label the parts of the plant and write what the plant needs to live and grow well. This activity will show you how well the students have understood the topic.

How well do you remember?

You may use the revision and consolidation activities on Student's Book page 14 as a paired class activity. If you are using the activities to assess individual learning, have the students work on their own to complete the tasks in writing. If you are using them as a class activity, you may prefer to let the students do the tasks orally. Circulate as they discuss the questions and observe the students carefully, to see who is confident and who is unsure of the concepts.

Some suggested answers
1 The wooden boat was once alive (as part of a tree). The bowl has never been alive.
2 Students' own lists
3 Both plants have leaves, flowers, a stem and roots. One plant has pink flowers and lots of stems. The other plant has one main stem and no flowers.
4 Students' own drawings; plants need light and water to survive

Consolidation

Consolidation and reinforcement of the students' understanding of the topic can be undertaken using Topic quiz sheets B1 and B2. These can be completed in class or as a homework task.

Topic quiz sheet answers

Sheet B1
1 Students' own drawings
2 Students' own drawings
3 The football is the odd one out because it is non-living.
4 False
5 Students' own answers

Sheet B2
1 Students' labelled drawings
2 Sweet potato; carrot
3 The middle plant is the odd one out because it does not have any flowers.
4 True; False

Biology • Topic 1 Plants Student's Book answers

Student's Book answers

Pages 2–3 (1.1)
1 Students' own descriptions; encourage discussion of the plants' appearance and size and of the different plant parts
2 The boy is watering the young tree to help it to grow.
3 Any suitable question such as: *Do plants need water to grow/survive? What will happen to plants that do not get water?*
4 The children can grow two plants and only water one, to see what happens to each plant.
5 Students' own predictions

Pages 4–5 (1.2)
1 Students' own answers
2 Living things might include: trees, people, flowers, grass, ducks, birds, goat.
3 Non-living things might include: fence, bench, buildings, food, clothes.
4 Elephant, tree, plant, goat, butterfly
5 Pencil, chair, bicycle, tea/mug, pot (for plant)

Pages 6–7 (1.3)
1 Living things might include: bird, animals, snake, spider, people, trees, pond plants, frogs, butterflies, duck, flamingos, tortoise.
2 Non-living things might include: fences, stones, rocks, teddy bear, glass, clothes, water, building, lamp post.
3 Plants might include: trees, pond plants. Animals might include: zebra, giraffe, antelope, snake, spider, duck, flamingos, frogs, butterflies, bird, tortoise, people.
4 five types: three types of tree, two types of pond plant
5 Same: they have green leaves, they are living, they are growing. Different: some live in water, some live on land, some are large trees, some are small plants.

Pages 8–9 (1.4)
1 Non-living things might include: fire, boat, hut, stones, river, mountains, cooking pot.
2 Any of the objects made from wood were once alive (trees).
3 The stones, mountains and water have never been alive.
4 A tree can grow and take in water because it is living. An object made from wood is non-living and cannot do these things.

Pages 10–11 (1.5)
1 They both have roots, green leaves, a stem and flowers. Only plant A has flowers.
2 Yes.
3 No.
4 Answers might include: yam, cassava, potato, swede, mooli.
5 Answers might include: the leaves of lettuces/cabbages, the seeds of fennel/coconut/sunflower.

Pages 12–13 (1.6)
1 They all have leaves, stems and flowers. Plants A and B are not healthy.
2 The unhealthy plants have not had enough water or light.
3 It is too hot and there is not enough water for the plants to grow well.

Biology • Topic 2 Humans and other animals

2.1 Parts of the human body

Student's Book pages 16–17

Biology learning objective
- *Structure and function*: 1Bs.03 Recognise and name the major external parts of the human body.

Thinking and working scientifically
- *Carrying out scientific enquiry*: 1TWSc.05 Collect and record observations and/or measurements by annotating images and completing simple tables.

Resources
- Workbook page 13
- PCM B4: Where do you wear it?
- Video B2: Children playing
- Digital resource B2: My body

Classroom equipment
- two balls, lining paper or wallpaper, coloured pens or pencils, scraps of fabric, glue
- recorded music that can be easily stopped and re-started

Key word
- body

 Supervise the students when they use glue. Make sure that the students take care not to jostle other students or bump into classroom furniture when they are performing movement-based activities.

Scientific background

The human body is made up of a head, a neck, a torso, two arms and two legs. The human body is designed to stand erect, to walk on two feet and to use the arms to carry and lift. The head has two eyes that allow us to see, two ears that allow us to hear, a nose that lets us smell and a mouth for eating and speaking. (Students will learn about sense organs and their five senses in Unit 2.2.) We also have hair to protect our scalp and eyebrows, and eyelashes to protect our eyes. The torso contains the majority of the body's organs, including the heart and lungs, which are located in the chest, and the digestive organs, which are located in the abdomen. The arms are joined to the torso by the shoulders, and there is a hand on each arm. Each hand has four fingers and an opposable thumb, so we are able to grasp things. The legs are joined to the torso at the hips. There is a foot on each leg, and five toes on each foot. All of the limbs are jointed, to allow flexibility and movement. At this stage, the students need to know the major external parts of the body.

Introduction

- Use the topic opener photograph on Student's Book page 15 as a talking point. Ask the students to describe what they can see. Ask them: *What do you think this topic will be about?* Let them briefly discuss this in groups. Tell them that they are going to learn more about humans.
- Say that you are going to draw something on the board and the students have to guess what it is.

Draw a body, bit by bit, on the board. Direct the students to Student's Book page 16. Discuss the picture and questions as a class. Ask: *Who knows what the parts of our body are called? Can you point to them on your own body?* Introduce the key word *body* and also the words for the body parts that are labelled in the picture: *head, shoulder, neck, arm, hand, finger, chest, stomach, leg, foot, toe.* Label the body you have drawn on the board.

- Make sure that the students all understand that these body parts are common to them all, and that although they may be of different sizes (e.g. leg length), they all serve the same function. Be sensitive when talking about the body parts in case not all children have all parts. Ask: *How many legs do you have? How many fingers do you have?*

- The students will have played body part games during their early years education. This orientation exercise gives the students something familiar to build on and will help increase their vocabulary. Play 'Simon says' with the class. If you say: *Simon says...*they must do the action that you say next. If you do not start with: *Simon says...*they should not do the action. Tell students to listen carefully and follow directions. Start with: *Simon says, put your finger on your head.* Give support by pointing to your head and putting your finger on it. Try: *Put your hand on your chest.* The students should ignore this.

Biology • Topic 2 Humans and other animals

Teaching and learning activities

- Ask the students to discuss, in their groups, all the things they can do. Make a class list on the board. Take feedback from the groups. Ask them to start their responses with: *I can…* Take an idea from each group, in turn, until they have no more suggestions.
- Stand in front of the students and hold up your hands. Indicate your feet, asking: *Why do we have hands and feet? What can they help us to do?* Arrange the students in two groups, each forming a large circle. Include an adult in each circle. Explain that the adult will name someone in the circle and then throw a ball to them. That student must try to catch the ball. Then they name another student and throw the ball to that student, and so on. Ask: *What parts of your body are you using to do this?* (arms, hands) Repeat, this time gently kicking the ball. Ask: *What parts of the body are you using now?* (legs, feet)
- Make sure students understand the questions on Student's Book page 17. Let them discuss their answers in groups. Take feedback as a class.
- Pin a long sheet of paper to a wall and ask a student to stand in front of it. Use a marker pen to draw round the outline of the student, to give a body outline. Repeat for each group. Pin each group's body outline to the wall and let the students add features to the face. Let them use fabric scraps to add clothes. Ask each group to add the labels at the correct places on the body outlines. Many at this stage will need help with the labelling, but the discussion of what the parts are is at least as important as whether they spell the words correctly.

Graded activities

1 Give each student a copy of PCM B4, which shows various items of clothing. Explain that they need to match each item of clothing to the body part that it is worn on. Students can work in pairs to do this. Circulate to check that they are matching correctly, asking questions to guide them as necessary. Once they have completed the activity, encourage them to discuss their experiences of wearing similar items of clothing.

2 Show the class Video B2, of some children playing. Ask: *What is happening? What parts of their bodies are the children using?* Let the students discuss this in groups, encouraging them to ask questions, share ideas and challenge the ideas of others. Take feedback and discuss answers as a class. Summarise by eliciting that the children are using all parts of their bodies.

3 Tell the students that you would like them to draw a picture of themselves on Workbook page 13. They should then label the different body parts. Circulate to check that they are adding the labels in the correct places. Ask questions such as: *Where is your arm? What is at the end of your arm? Can you find this on your drawing?* This activity should show you how well the students have understood the unit.

Consolidate and review

- Play 'Musical statues.' Explain that when the music is playing, students can move about and play; as soon as the music stops, the students must 'freeze', standing still, just like a statue. If a student moves, say: *I saw you moving your arm/leg/head, you are out of the game.*
- Students discuss what they know about hands and what they can do with them. Ask them to demonstrate, for example, wiggling their fingers.
- Let the students complete Digital resource B2 to consolidate their learning.

Differentiation

■ All of the students should be able to match the clothing to the correct body parts. Even if the clothing is unfamiliar to the students, they should still be able to match correctly using the size and shape of the clothing.

● Most of the students should be able to talk and work together to share ideas and identify the different body parts. Some students may discuss that different body parts can do different things.

▲ Some of the students should be able to draw an accurate representation of themselves and correctly label the different body parts. Some may struggle with the labelling; if so, offer support and guidance.

Biology • Topic 2 Humans and other animals

2.2 Our senses

Student's Book pages 18–19

Biology learning objective

- *Structure and function:* 1Bs.02 Identify the senses (limited to sight, hearing, taste, smell and touch) and what they detect, linking each to the correct body part.

Thinking and working scientifically

- *Carrying out scientific enquiry:* 1TWSc.04 Follow instructions safely when doing practical work; 1TWSc.05 Collect and record observations and/or measurements by annotating images and completing simple tables.

Resources

- Workbook pages 14 and 15
- Slideshow B5: Colours
- Audio clip B1: Animal sounds
- PCM B5: Fruity smells
- PCM B6: Senses in action
- PCM B7: Whose eyes?

Classroom equipment

- mystery objects, a soft fabric bag
- empty drink cans, sticky tape, rice, sand, small nails, tacks
- selection of different objects, white sheet
- four small pots with lids, four different fruits with a strong smell

Note: Before the lesson, prepare four small plastic pots. Label them from 1 to 4, and make some small holes in the lid of each one.

Key words

- sense organ • senses

 Be aware that some students, particularly boys, may be colour-blind and have problems differentiating certain colours.

Scientific background

The human head has on it all of our *sense organs*: two *eyes* for *sight*, two *ears* for *hearing*, a *nose* for *smell*, a *tongue* for *taste* and *skin* for *touch*. The sense organs are connected by nerves to the brain, which tells us the things that our senses detect.

Although we have five senses, we rarely use them individually. For example, we use smell and taste together, and often support one sense with another. If we lose one of our senses, it may be that our other senses become sharper, to compensate.

Introduction

- Draw a face on the board. Point to each of the five sense organs and ask: *What is this? What do we call this part of the face?* Elicit the correct names and label the face.
- Explain that some parts of our faces have important jobs. Ask: *What are these jobs? What do different parts of our faces do? What are our eyes, ears, nose and mouth for?* Let the students discuss their answers in small groups. Take feedback, discuss their ideas and explain that eyes let us see, ears let us hear, our nose lets us smell and our mouth lets us talk and eat.

- Show Slideshow B5, about colours. For each colour, ask: *What is this colour?* On the last slide, point to colours at random and ask: *What colour is this?* Concentrate on the colours that were unfamiliar to the students. Ask: *Which part of your body do you use to see the colours?* Explain that the students are using their eyes to see.
- Introduce the term *sense*. Many students will confuse the meaning with common usage, e.g. *They do not have any common sense.* Explain to them that in the context of their body the term has a very specific meaning. Introduce the five sense key words and write each one on the board next to the correct sense organ: *eye – sight*, *ear – hearing*, *nose – smell*, *tongue – taste*, *skin – touch*. Point out the spelling and pronunciation of all the words for this unit. Ask the students to repeat each word after you.

Teaching and learning activities

- Make sure students understand the questions on Student's Book pages 18 and 19. Talk about what is happening in the pictures. Ask: *What can the girl smell? Does she like the smell? Do you like the smell of flowers?* Ask: *What can the boys*

Biology • Topic 2 Humans and other animals

taste? Which sense are they using? Point out that we often use more than one sense at a time.

- Put equal amounts of a substance into a pair of empty drink cans. Repeat for a variety of substances, such as rice, sand, small nails, tacks, water and polystyrene packing beans, to produce a set of pairs of cans. Tape the openings shut. Shuffle the cans. Ask the students to match the cans into pairs that contain the same substance by shaking them and comparing the sounds. Ask: *What made you sort the cans in this way?* Establish that they used their ears and that hearing is one of our senses.

- Show the class a soft fabric bag with something hidden inside it. Invite individual students to feel the object, through the bag, and describe what they feel. Ask the students if the object is rough, smooth, hard or soft, and so on. Ask the students to guess what the object is. When they make a correct guess, repeat with a different object. Establish that we use our hands to feel things and our sense of touch to identify them.

- Play a game with about eight objects placed on a white sheet. Cover them over after about 15 seconds and ask the students to tell you what was on the sheet. This can be made progressively harder by adding more objects, to demonstrate that the eye is the organ of sight and the brain has to remember what the eye has seen.

- Put a slice of fruit into each small, numbered pot. Show the group that you can smell the fruit through the holes in the lids. Then let the students smell the four pots and try to identify the fruit inside. Ask them to draw a picture of the fruit that they have identified, in the table on PCM B5. When the students have completed the task, take feedback and see how many students guessed correctly. Students can add the correct names of the fruit to the table. Ask: *Was it more difficult to identify the fruit when you could only use your sense of smell?* Stress again that we often use more than one sense at a time.

Graded activities

1 Let the students complete the activity on Workbook page 14. They should match each sense to the correct sense organ. Circulate, asking questions to guide their thinking: *Which part of our body do we use to see? When we see something, which sense are we using?*

2 Let the students complete the activity on PCM B6 in pairs. They need to identify the senses that each person is using and then explain why. Ask questions such as: *What is this person doing? What senses will they need to do this? Why do they use these senses? Will the person washing their hands just be using the sense of touch?* (No, they will also be smelling the soap, etc.) The students should recognise that some people may be using all of their senses at the same time, while others will be using different combinations. Remind the class which sense organ detects which sense, and how that sense can help to protect us.

3 Let the students complete the activity on Workbook page 15. They should draw lines to label the picture and then complete the sentences to describe what each sense does. Ask questions such as: *What job does the nose do? What job do your eyes do?*

Consolidate and review

- Ask the students to listen to the range of animal sounds on Audio clip B1. Ask: *Could you tell what each one was?* (tiger, gibbons, hawk, donkey, cricket and bird)

- Explain to students that just like humans, animals have senses and use their sense organs to detect changes in their environment. Elicit, through questioning, that in many animals, some senses are more developed than others. For example, animals that hunt at night may have a very highly developed sense of smell. Some animals can see things that are very far away, while others can see well in the dark. Student's can complete PCM B7 for enrichment purposes.

Differentiation

■ All of the students should be able to correctly match the senses to the sense organs.

● Most of the students should be able to identify the senses that the different people are using. You may need to remind students that it is possible to use more than one sense at a time. Some students may need help in explaining the reasoning. If so, circulate to offer support and guidance as necessary.

▲ Some of the students should be able to label the picture correctly and complete the sentences with little prompting.

Biology • Topic 2 Humans and other animals

2.3 Using our senses

Student's Book pages 20–21
Biology learning objective
- *Structure and function:* 1Bs.02 Identify the senses (limited to sight, hearing, taste, smell and touch) and what they detect, linking each to the correct body part.

Thinking and working scientifically
- *Carrying out scientific enquiry:* 1TWSc.05 Collect and record observations and/or measurements by annotating images and completing simple tables.

Resources
- Workbook pages 16 and 17
- Slideshow B5: Colours
- Audio clip B2: Familiar sounds
- Audio clip B3: Unfamiliar sounds
- Digital resource B3: Senses

Classroom equipment
- soft toys
- small plastic containers with small holes in their lids
- cotton-wool balls
- samples of strongly scented substances, such as lemon, orange, vinegar, mint, garlic
- large sheets of paper, coloured pens or pencils
- selection of items for sensory hunt

Key words
- senses • danger • safe

 Check if any of the students have food allergies. Never give them nuts. The students must not eat anything unless they are told that they may. Whenever students are required to taste things in the classroom, hygiene precautions need to be taken. If you take the students on a 'sensory hunt' outdoors, ensure they are safe and that they stay together.

Scientific background

Humans and animals have five senses: *touch*, *taste*, *smell*, *sight* and *sound*. These are detected by the sense organs: *skin*, *tongue*, *nose*, *eyes* and *ears*. We use our senses to gather information about our environment and to protect us from danger. Sound travels in waves through the air and into our ears, and causes the eardrum to vibrate. The eye works like a camera: it takes in light from whatever we are looking at and makes a tiny picture of it on the back of the eyeball. When we breathe, air goes into the nose through the nostrils. The tongue is covered with about 10 000 taste buds. Each sense organ is connected by nerves to the brain, which tells us the things that our senses detect. The nervous system allows the body to respond to changes in the environment. These responses are usually controlled by the brain and are informed by our senses, which is how they help to keep us safe.

At this stage, students do not need to understand the detail of how the body works, but should know that our senses tell us about the world around us and can help keep us from danger.

Introduction
- Ask: *Who can remember all the colours from the previous lesson?* Use the final slide in Slideshow B5 to review the colours. Ask: *What did you eat for your evening meal yesterday? What colours were on your plate? What did the different foods taste like?*
- Ask the students to tell you the names of all their senses. If necessary, review the sense organs and the names of the senses. Ask: *What things do you like/not like to touch? Why do you like the feel of some things?* Repeat the questions for the other senses.
- Ask the students to look at the picture on Student's Book page 20. Talk about the picture and answer the questions as a class.

Teaching and learning activities
- Ask the students to use their hands to touch something, such as a soft toy. Next, tell them to touch it with their nose, then their cheeks. Ask: *Can you still feel it?* Let them try with their elbows. Say: *You can feel things with your skin all over your body, but some places, especially*

Biology • Topic 2 Humans and other animals

your hands and fingers, can feel things more. Ask the students: *What else can you tell by using your sense of touch?* (hot and cold things, rough and smooth things, hard and soft things)

- Explain that the brain acts very quickly. Say: *If you touch a hot stove it will hurt you, and you will move your hand immediately. This is one way our sense of touch protects us.*

- Put a damp cotton-wool ball into each of the prepared containers. Sprinkle samples of strongly scented substances over the wet cotton-wool balls. (The damp enhances the scent.) Use the same substance in two containers, but make all the rest different. Put the lids on the containers. Ask the students to identify as many scents as they can. They should try to match the two that are the same. Ask: *Were you correct? Was it easy to identify the smells?* Ask the students to suggest words to describe some smells. Take feedback and write the descriptive words on the board for the students to refer to during the lesson. Include *strong*, *weak*, *pleasant*, *unpleasant* as well as more specific terms.

- Direct the students to Student's Book page 21. Let them discuss, in small groups, what they see in the picture. Ask: *What will people see on the street? What sounds will there be? What will the people use to detect danger?* Elicit how our different senses can help to keep us safe.

Graded activities

1 Play Audio clip B2 of eight familiar sounds. Ask the students to identify them. The students should write their answers in the table on Workbook page 16. Then play some unfamiliar sounds from Audio clip B3. Ask the students to describe the sounds, even though they do not know what they are. (fire burning, dolphin calls, keyboard on a computer, a pan of water boiling) Ask: *Is the noise loud or quiet? Is it an animal? Is it a machine? Is it easy to identify sounds when you cannot see what is making them?* Establish that the ear is the organ which detects sounds, and that some sounds are more familiar than others.

2 Put the students in pairs. Ask each pair to choose one of the senses and make a poster to describe that sense and how it helps in our daily lives. Let the pairs take turns to present their posters to the class. Ask: *Why do we have senses? What would it be like if we couldn't see, or hear, or feel?* Tell the students that humans who were born without, or who have lost, one of their senses have been found to have developed their other senses more. For example, a blind person's senses of hearing and smell may be heightened. Ask the students to tell you why they think this happens. Remind the class of how important our senses are for survival.

3 Let the students undertake a 'sensory hunt', inside the classroom or outside. For an indoor activity, add a variety of objects not normally found in the room, for example something that is orange (or another colour), something sweet, something that makes a noise, something with a rough texture and something with an odour. Students should make a list of everything they detect with their senses. Back in the classroom, ask: *Can some objects fit into more than one 'sense' category? Which sense did you use the most? Did you use more than one sense at any time?* Ask students to choose one place they visited and to draw a picture of what their different senses detected there.

Consolidate and review

- Use Workbook page 17 to consolidate the teaching in this unit.
- Let the students complete Digital resource B3 to consolidate their learning.

Differentiation

■ All of the students should be able to identify a range of familiar and unfamiliar sounds. If some students only know the names in their own language, provide them with the English translations.

● Most of the students should be able to work collaboratively to share ideas and think critically to produce an informative poster about one of the five senses. Most should be able to do this with little help. For those that need some help, you could provide them with a suggested frame on the board, or ask questions to guide their thinking.

▲ Some of the students should be able to list a range of familiar things which their senses detected, such as the sound of other students. A few students may identify less familiar things, such as the feel of the surface they are walking on.

Biology • Topic 2 Humans and other animals *Science in context*

2.4 Users of science

Student's Book pages 22–23

Biology learning objective
- *Structure and function*: 1Bs.02 Identify the senses (limited to sight, hearing, taste, smell and touch) and what they detect, linking each to the correct body part.

Science in context skills
- 1SIC.03 Know that everyone uses science and identify people who use science professionally.

Resources
- Workbook pages 18, 19 and 20

Classroom equipment
- large sheets of paper
- coloured pens and pencils

Key word
- sense organ

 Make sure that students take care not to bump into each other or into classroom furniture when they are doing movement-based or blindfold activities. If the students use the internet, ensure they do so safely and always under adult supervision.

Scientific background

We all apply science in everyday life, often without realising it; for example, chemical changes in cooking and physical forces when playing in a park. There are many professions that use science, from doctors and pilots to farmers and builders. This unit will put science in context by focusing on our sense of sight and the role that ophthalmologists and opticians play in helping to look after our sight.

The human eye is a complex organ. The white part of the eye is called the sclera, the coloured part is called the iris and the pupil is the black centre of the eye. When light enters the pupil, it hits the lens. The lens focuses light on the retina at the back of the eye. The retina then changes the light into nerve signals, which it sends to the brain so that the brain can understand what the eye is seeing.

Ophthalmologists and opticians are professionals who help care for our sight. When we visit an optician to have an eye test, they carefully check the different parts of our eyes. If any parts are not working well, this can affect our vision. An optician can correct this by giving us glasses or contact lenses, which help us to see more clearly.

Students do not need to understand the detail of how the eye works but should appreciate that our sense of sight is important and that opticians can help us to protect this sense. This unit will get students to start thinking about the use of science in real-world contexts.

Introduction

- Ask: *Who can remember the names of our five senses? Which sense organ do we use for each sense?* Check that students can recall these correctly.
- Ask students to look at the picture on Student's Book page 22. Answer the first two questions as a class. Then get the students to work in pairs to think of things they do with their sense of sight. If necessary, prompt them with questions such as: *What things can you see/watch at home? Do your eyes help you to do this? What things do you do at school? How do your eyes help you to do these things?* Take feedback and write students' ideas on the board.
- Demonstrate the importance of sight with two simple experiments. First, ask students to cover one eye and look at a book. Ask: *Do you find it harder to read like this? Why?*
- Next, ask students to stand on one leg with their eyes open. Then ask them to do it with their eyes shut (without holding onto anything). Ask: *Do you notice any difference?* Take feedback as a class.

Teaching and learning activities

- Ask students: *Is our sight important? Are there some things we cannot do without our sight? Do we use sight only for fun or does it also help us?* Discuss as a class. Be sensitive to any students who may wear glasses or have sight problems. Guide students to understand the importance of sight in helping us with everyday tasks and in

Biology • Topic 2 Humans and other animals

keeping us safe. Tell the class that they are going to learn about a person whose job is to help care for our sight.

- Ask students to look at Student's Book page 23. Talk about the pictures and answer the first question as a class. Then get students to work in small groups to share any experiences they have of visiting an optician. Tell the class about your own experiences and then invite students to share theirs with the class.
- Ensure that students understand the role of an optician. Explain that an optician will test your eyes and, if you cannot see clearly, they can help by giving you glasses. Address any misconceptions. Reassure students that an eye test is nothing to worry about and that wearing glasses is very common.
- Ask students how an optician can check how well a person can read things far away. Guide them to the concept of an eye test chart and tell students that they are going to work together to create their own chart. Give each pair a large sheet of paper and some coloured pens or pencils. Ask them to design a chart with larger items at the top (which are easier to see) and smaller items at the bottom. Direct them to the picture on Student's Book page 23 as an example, but tell them that they can use letters, numbers, pictures or any combination of these – they can be as creative as they like. If some students struggle with ideas, put some examples on the board to help them.
- When students have completed their charts, they can work with another pair to role play being an optician and give each other an eye test. Again, ensure sensitivity to students who may have sight problems, to avoid any distress or upset.

Graded activities

1 Let the students complete the activity on Workbook page 18. They should circle all the things that we primarily use our sense of sight to enjoy. Circulate to check understanding and ask questions to guide their thinking, if necessary.

2 Tell students that they are going to work in pairs to try doing two everyday tasks while wearing a blindfold. Give them some tasks to choose from, such as writing their name, drawing a shape, putting on a jacket, tying their shoelace, etc. Ask them to draw and write what they expect to find on Workbook page 19. To help them do this, circulate asking: *Will you be able to see? Will this make it harder to do?* Students should then attempt the tasks and record what they found out on Workbook page 20. Invite pairs to present their findings to the class. Encourage students to use as much detail as possible. To finish, they should decide whether their findings matched their predictions or not.

3 Working in groups, help students to do some research using reference books or the internet. Students should make a fact sheet to summarise the unit. Display these on the wall in the classroom and discuss.

Consolidate and review

- As a class, recap the things that our eyes help us to do and how they can keep us safe. Refer to the students' earlier ideas on the board and try to add some more. Elicit why it is important to look after our eyes. Discuss some of the things we can do to protect our eyes (e.g. visit an optician, wear sunglasses, wear protective goggles).
- Let the students work in groups to discuss what they have learned about the role of an optician. Encourage them to each form a sentence. Take feedback as a class.
- Ask students to discuss, in groups, other jobs that use science, such as scientists, doctors, etc. Invite groups to share their ideas with the class and discuss.

Differentiation

■ All of the students should be able to identify the things that we use our sense of sight for. If not, help the students by reminding them of some things that our sight helps us to see.

● Most of the students should be able to say what they expect to happen and then work collaboratively to attempt the tasks. Most of the students should be able to draw or write their findings with some support.

▲ Some of the students should be able to work collaboratively, thinking critically and asking sensible questions to extend their knowledge and produce an informative fact sheet. Some students may not be working at this level yet, so you may prefer to do this as a more structured task by offering a frame to work to.

Biology • Topic 2 Humans and other animals

2.5 What do animals need to survive?

Student's Book pages 24–25

Biology learning objective
- *Life processes:* 1Bp.02 Know that animals, including humans, need air, water and suitable food to survive.

Thinking and working scientifically
- *Carrying out scientific enquiry:* 1TWSc.01 Sort and group objects, materials and living things based on observations of the similarities and differences between them.

Resources
- Workbook page 21
- PCM B8: Spot the differences
- PCM B9: My family meals
- PCM B10: 1–4 spinner
- PCM B11: Race to the water hole
- PCM B12: Animal spinner

Key words
- air • water • food

Classroom equipment
- clean, empty pet-food packaging
- bottles or cups of clean water
- bottles or cups of dirty water (water mixed with mud, leaves, twigs, etc.)
- large sheets of paper, coloured pens or pencils

 Do not let students drink any of the water that they handle. Mop up any spills immediately. If the students use the internet, ensure they do so safely and always under adult supervision.

Scientific background

Humans and other animals need air, water and suitable food to survive. Food provides animals with the chemicals they need for growth and fuel for energy, and it helps to protect them from illnesses. Animals cannot make their own food and need to eat plants and other animals to survive. Every animal, including humans, has a particular diet that it needs in order to stay healthy. Carnivores eat only meat and herbivores eat only plants. An omnivore eats both meat and plants. It is important for humans to eat a variety of different food types and that we eat the right amount of each type of food.

Humans and other animals need to drink water to stay alive, for their bodies to function well and to be healthy. Water from most sources must be cleaned before we can drink it. Clean water is safe to drink but dirty water can make humans and animals ill, and in extreme cases can lead to death. Fresh water comes from a wide variety of sources, such as rivers, lakes and wells, as well as taps and bottles. Seawater has salt in it so is not good to drink.

At Stage 1 the students do not need to understand the detail of diet or different food groups, but they should recognise the importance of food and water for animals to stay alive and be healthy.

Introduction

- Ask: *Does anyone have a pet? What food does your pet eat? Do all animals eat the same kinds of food?* Take feedback. Elicit that some animals eat only meat (for example, lions, frogs, eagles), some animals eat only plants (for example, goats, butterflies, dugongs) and some animals, including humans, can eat both. Discuss as a class.
- Give students some clean, empty pet-food packaging or print images from the internet. Ask students to look at the labels to see what the food contains. Ask questions such as: *Is there meat in this food? Does it contain any plants?* Establish that some pets eat meat while others eat only plants.

Teaching and learning activities

- Ask the students to look at the pictures on Student's Book page 24 and answer the questions as a class. As you take answers, create three lists of the board: one for animals that eat only meat, one for animals that eat only plants, and one for animals that eat both. Ask the students to name some more animals to add to the lists. Help them to identify and sort animals into the correct groups.

Biology • Topic 2 Humans and other animals

- Ask: *Do all animals need to eat? Why?* Take feedback. Establish that humans and other animals need food to grow and stay healthy.
- Give the students a few minutes to think about times when they have been hungry. Ask: *What does it feel like if you haven't eaten for several hours? What might happen to you if you did not eat food for a very long time? What can you do to make you feel better when you are hungry?* Discuss the fact that, if animals do not have enough food, they can become unwell (or even die). Eating food restores their energy and makes them feel well again.
- Refer students to the pictures on Student's Book pages 24 and 25. Ask: *What else do animals need? What is the boy doing?* Take feedback as a class and elicit that all living things, including animals, need water to stay alive and be healthy. Ask: *What would happen if you did not drink for a couple of days?* Allow them to discuss this in groups and then gather their ideas. Students should be able to appreciate that water, as well as food, is essential for life.
- Give each pair of students a bottle or cup containing some clean water and another containing some dirty water. Let them examine both. Say: *Which one would you drink if you were thirsty?* Elicit that it would be the clean water. Ask the students to give reasons for their choice.
- Explain that water from most sources must be cleaned before it can be drunk or it can make you ill. Explain that there is a difference between fresh drinking water and the water in the sea, which is salty and not good to drink.
- Make sure students understand the questions on Student's Book page 25. Discuss the answers as a class. Point out the picture of the well and make sure that the students know what it is and how it is used. Give each student a copy of PCM B8 and ask them to circle the five differences.

Graded activities

1 Let the students complete the activity on Workbook page 21. They should tick the sources of water that are safe to drink. Circulate, asking questions to guide their thinking: *Do you think the water in the river is clean? If it is not clean, is it safe to drink?*

2 Ask the students to think about food that they regularly eat with their family and to draw pictures on a copy of PCM B9. They should label the pictures, in their own language or in English. Depending on ability, some students may add one label for each complete meal while others can label the individual food items. When they have completed their drawings, let the students discuss their meals in groups. Take feedback as a class. Ask: *What is your favourite food?* Compile a class list on the board of some of their favourite foods.

3 Explain that what an animal eats often depends on the part of the world that it lives in and what types of food are available locally. Ask the students if they know any animals that live in other countries. Write some examples on the board. Next, ask the students to choose a country and two animals that they would like to find out more about. Working in groups, help them to do some research using reference books or the internet. Students should make a poster to illustrate their findings. Display these on the wall in the classroom and discuss.

Consolidate and review

- Give the students a number spinner made from PCM B10. They take turns to spin the spinner and move their counter on the board game on PCM B11. The students are gazelles and must reach the waterhole without being eaten by lions or cheetahs.
- Give the students an animal spinner made from PCM B12. They take turns to spin the spinner, identify the animal and say whether it eats plants or meat.

Differentiation

■ All of the students should be able to identify which sources of water are safe to drink. Circulate, asking questions, if the students need some guidance.

● Most of the students should be able to draw an accurate picture of their chosen meals and add labels with little help.

▲ Some of the students should be able to work collaboratively, thinking critically and asking sensible questions, to extend their knowledge and produce an informative poster. Some students may not be working at this level yet; if so, you may prefer to do this as a more structured activity by offering a frame for them to work to.

Biology • Topic 2 Humans and other animals

2.6 Humans are similar

Student's Book pages 26–27

Biology learning objective
- *Life processes:* 1Bp.04 Describe how humans are similar to and different from each other.

Thinking and working scientifically
- *Carrying out scientific enquiry:* 1TWSc.01 Sort and group objects, materials and living things based on observations of the similarities and differences between them.

Resources
- PCM B13: My family

Classroom equipment
- old magazines
- scissors and glue
- paper
- colouring pens or pencils
- students' photographs from home of themselves and of a family member
- large sheets of card

Key words
- human • similar • feature

 Supervise the students when they are cutting with scissors and working with glue. Teachers should be aware of the need to show sensitivity to students who may be adopted or not in touch with their biological family for other reasons.

Scientific background

All human beings have some similar features. People have the same general sort of shape, same body parts and same facial features, but the appearance of these and characteristics such as height, weight and hair colour vary from person to person. During reproduction, parents pass on genetic material to their children. However, not all the characteristics a parent can offer are passed on. Therefore, children inherit features from both parents. These features are called *hereditary characteristics*, and include things such as hair and eye colour, and whether the tongue can be rolled or not. Some characteristics such as scars are acquired during a lifetime rather than inherited. Because of genetics, people within a family can bear a very strong likeness to each other. However, even people from very different ethnic groups have recognisable features that are the same. In this unit, the students will focus on similar features that people share. (Differences between people will be covered in Unit 2.7.)

Introduction

- Ask the class to look at the picture at the bottom of Student's Book page 26. Ask: *What can you see in the picture?* Take feedback and guide the students to establish that the photo shows an extended family: three generations of the same family. Ask: *Can you see any things that are the same?* Elicit that people in a family often have some things in common. If necessary, recap the major external parts of the human body covered in Unit 2.1.

- Introduce the key words: *human, similar, feature*. Write them on the board and point out how each word is spelled and pronounced. Ask the students to repeat the words after you. Explain that we use the term 'similar features' to describe things that look the same between different people. Ask the students to think of a similar feature that they have in common with someone else. For example, ask: *Do you have brown hair? Who else has brown hair?* Introduce new words, as necessary.

Teaching and learning activities

- Ask the students to brainstorm some similarities between humans. Remind them to think about people from different countries around the world, not just the people they know personally. Allow them time to discuss their ideas in groups. Then take feedback and compile a class list on the board.

- Encourage the students to bring into school a photograph of themselves. Mix up the photographs and ask one student to pick and look at one photo (make sure no one sees which picture they have selected). They must describe the student in the photograph without using their name. The other members of the class should guess who it is from the description. Be sensitive to the students in your

Biology • Topic 2 Humans and other animals

class to avoid embarrassing any of them. Shuffle the photographs again and repeat for another member of the class. When a number of rounds have been played, the students will be familiar with the process of describing someone using fine detail. The students can use this skill in the graded activities that follow.

- Have the students work in pairs to draw each other and colour their pictures. Ask them to discuss how they are similar and how they are different. (Differences between humans will be covered in the next unit.) Take feedback, encouraging the students to use a wide range of adjectives, e.g. *tall*, *short*, *brown*, *blue*, etc., and comparatives, e.g. *bigger*, *smaller*. If some students need support, you could write a list of suitable words on the board to give them ideas.

- Make sure the students understand the question on Student's Book page 27. Ask them to find and name the features that the children have in common. Let them discuss their answer in groups. If the students need some guidance, refer them to the class lists on the board. This activity will reinforce the idea that, although people come from many different countries and cultures around the world, they all have some features in common.

Graded activities

1 The students should collect as many pictures of different people as possible. These can be real family photos or cut from magazines that you bring to class. Ask groups of students to sort a selection of pictures according to their own criteria. They should then explain their reasons to the rest of the class. Ask: *Why did you sort the photos in this way? What features does each group have in common?* The criteria could include age, hair colour, height, etc. The choice of criteria is not important, but students should correctly sort the pictures so that all photos in a given group fulfil the chosen criteria.

2 Teachers should be aware of the need to show sensitivity to students who may be adopted or not in touch with their biological family for other reasons. Give each student a copy of PCM B13. Ask them to think about members of their family and the similar features that they share with them. They should draw pictures and write the names of the family members on PCM B12. Circulate, offering support by asking: *Which things are the same? Are your ears similar? Do you have the same colour eyes?*

3 Teachers should be aware of the need to show sensitivity to students who may be adopted or not in touch with their biological family for other reasons. Tell the students that they are going to do a short presentation about someone in their family. Ask the students to bring into school a photograph of a family member. Ideally, they should share some similar features. Allow the students time to prepare. Circulate, offering support and encouragement. Remind the students to describe the family member in as much detail as possible and to point out the features that they have in common. Then divide the class into groups and ask each student to give their presentation to the rest of their group.

Consolidate and review

- Give each student their photograph, mounted on a large sheet of card. Ask the students to make a poster about themselves. They should write their name, their age and some words to describe their appearance.

- Ask the students to make up a short story about their chosen family member from the activity. Let them tell their story to classmates in their group.

Differentiation

■ All of the students should be able to identify some similar features between people and correctly sort the pictures into groups based on criteria of their choice. Most of the students should be able to explain the reasoning for their chosen criteria with some support.

● Most of the students should be able to draw pictures of family members and describe some similar features. Some may find the descriptions harder than others. Circulate, offering support and asking questions to guide their thinking.

▲ Some of the students should be able to give a detailed description of shared features with a family member of their choice. Some may be able to do this independently, while others may need more help; if so, you could write a frame on the board to provide a structure for their presentation.

Biology • Topic 2 Humans and other animals

2.7 Humans are different

Student's Book pages 28–29

Biology learning objective
- *Life processes:* 1Bp.04 Describe how humans are similar to and different from each other.

Thinking and working scientifically
- *Scientific enquiry: purpose and planning:* 1TWSp.02 Make predictions about what they think will happen.
- *Carrying out scientific enquiry:* 1TWSc.03 Take measurements in non-standard units; 1TWSc.04 Follow instructions safely when doing practical work.
- *Scientific enquiry: analysis, evaluation and conclusions:* 1TWSa.01 Describe what happened during an enquiry and if it matched their predictions.

Resources
- Workbook page 22
- PCM B14: Picture of my friend
- PCM B15: Spot the differences
- Slideshow B6: What can you do?

Classroom equipment
- rulers, squared paper
- paper, washable stamp pads or non-toxic paint, sponges, magnifying hand lenses

Key words
- different • unique

 The students must thoroughly wash their hands after making their fingerprints.

Scientific background

All living things vary in many ways; in fact, no two are alike. Even identical twins are different in certain ways. Humans all have the same general sort of shape, but their height, weight, face shape, etc., are all different.

Growth of the young human body takes place continuously, although the process takes place so slowly that we do not usually notice the changes. Some children grow faster than others, though all children go through periods of differing growth rate. As we grow, all parts of our body grow bigger until we reach adulthood. At this point most growth stops, apart from hair and nails. Various measurements can be used to measure human growth accurately: these include height, hand span and foot size. Weight is not a good measurement to use, as it is greatly affected by what and how much we eat. As we grow, our bodies change and we learn new skills. Children learn to do things for themselves as they grow and develop.

Introduction

- Remind students that all people have the same sort of bodies, with a head, arms and legs, but explain that they also have many things which are different, such as height, hair colour, eye colour, etc. Ask the class to look at the top picture on Student's Book page 28. Ask: *In what ways are the children the same? In what ways are they different?*
- Introduce the key words for this unit: *unique* and *different*. Write the words on the board and point out how each one is spelled and pronounced.
- Say: *We may have some features in common, but we also have differences.* Explain to the class that everyone's fingerprints are different. Give each group a washable stamp pad or non-toxic paint, a sponge and paper, to make fingerprints. Let them examine each other's fingerprints to check that they are different. Provide magnifying hand lenses so they can look closely. Ask: *What other things about people are different?* Take feedback.

Teaching and learning activities

- Ask the students to discuss some of the ways the body changes as it grows. Use these prompts: *Say how your body changes as you grow. Will you keep growing forever?* Make a list of the ways in which humans change as they get older.
- Ask the students to discuss how they know that they have grown. They should be aware they have grown when clothes no longer fit, or when their hair or fingernails need cutting. Discuss how we can measure the growth of a child and how

Biology • Topic 2 Humans and other animals

different parts of the body change as we grow. Ask: *What parts of your body could you use to measure how you grow over a period of time?* Take feedback and discuss their ideas. Tell the students that height is not the only feature that shows growth.

- Mark a line along the ground, about 1.5 metres long. Show the students how to measure it in 'foot lengths', starting with the heel of one foot at one end, then placing the heel of the other foot immediately in front of the first foot. Provide a similar line for each group. Let them count and record how many 'steps' each student takes to reach the end of the line. Ask them to compare their results.

- Ask the students if they think that the older students in the class are the tallest. Take a class vote on it and record what the students think. Ask the students to line up. Arrange them into height order, from tallest to shortest. Next, ask the students to form another line, but this time in age order, from oldest to youngest. Ask: *Are you standing in the same place as before?* Ask the students if their predictions were correct. It is unlikely that you will find a pattern of oldest to youngest giving a direct correlation to height. This is a good point for discussion, and the students should try to explain why a pattern was not necessarily found. Ask: *Why was the tallest child not the oldest?* (Because children grow at different rates.)

- Make sure students understand the questions on Student's Book page 29. Let them discuss their answers in groups. Then take feedback as a class.

Graded activities

1 Ask the students to work in pairs. Give each student a copy of PCM B14. Explain that they need to draw a picture and describe their partner's features. Once the students have completed the activity, ask them to say how they are different from each other. The rest of the class can comment on whether they agree or not.

2 Discuss the different things that people can do at different ages. Ask: *What can you do now that you could not do when you were two years old?* Show the class Slideshow B6. Ask: *What can you do now that you could not do when you could only crawl? What can you do that a grandparent cannot? What can an older brother or sister do that you are not allowed to do? Allow students to discuss their ideas in groups.*

3 Tell the students that they are going to investigate and compare the size of their hand spans. Ask: *Do you think that the oldest student in the class will have the biggest hand span?* As a class, discuss how to take fair measurements. Ask: *Where on the hand are you going to measure the hand span? What will you use to measure?* Some students may prefer to draw around their hands on squared paper and compare the drawings. Most students should be able to measure using a ruler. Ask students to complete the activity on Workbook page 22. Discuss their findings as a class. Ask whether their predictions were correct and for them to try to explain why.

Consolidate and review

- Give each student a copy of PCM B15. Ask them to look at the two faces and to find and circle the five differences.

- Ask questions to encourage a class discussion: *Who is the tallest person in the class? Did they have the biggest hand span? What do you think affects how fast we grow?* (age, health and diet)

Differentiation

■ All of the students should be able to draw a reasonable picture of their partner from first-hand observations, describe features and suggest a range of differences when comparing.

● Most of the students should be able to work in groups to share ideas when identifying different abilities at different stages of growth and development. Students working at a higher level may start to describe physical characteristics as a way of explaining the differences.

▲ Some of the students should be able to make predictions and then follow instructions to take accurate measurements. Some may need more help than others. Circulate, offering help with completing the practical element if necessary and then asking questions to lead their thinking when comparing their results.

Biology • Topic 2 Humans and other animals Consolidation

Consolidation

Student's Book page 30

Biology learning objectives
- *Structure and function:* 1Bs.02 Identify the senses (limited to sight, hearing, taste, smell and touch) and what they detect, linking each to the correct body part; 1Bs.03 Recognise and name the major external parts of the human body.
- *Life processes:* 1Bp.02 Know that animals, including humans, need air, water and suitable food to survive; 1Bp.04 Describe how humans are similar to and different from each other.

Resources
- Workbook page 23
- Topic quiz sheets B3, B4 and B5

Looking back Topic 2

- Use the summary points on Student's Book page 30 to review the key things that the students have learned in the topic. Ask questions such as: *In what ways are you similar to your classmates? In what ways are you different? Can you name the different parts of your body? What do animals need to survive? Where can we find clean water to drink? What do your eyes, ears, nose, tongue and skin help you to do? Describe how you are different now from when you were a baby.*
- Ask the students to look at the picture on Workbook page 23. Discuss it as a class. Ask: *Have you ever been to a fun fair or festival?* Tell them to imagine they are at the fair. Ask: *What would you see? What would you hear? What would you feel? What would you smell? What would you taste?* Allow them time to discuss this in groups. Then ask the students to write what the children in the picture can see, hear and smell at the fun fair. This activity will show you how well the students have understood the topic.

How well do you remember?

You may use the revision and consolidation activities on Student's Book page 30 as a paired class activity. If you are using the activities to assess individual learning, have the students work on their own to complete the tasks in writing. If you are using them as a class activity, you may prefer to let the students do the tasks orally. Circulate as they discuss the questions and observe the students carefully, to see who is confident and who is unsure of the concepts.

Some suggested answers

1. The children can see other people, fairground rides, food stalls, balloons. They can hear music, people talking and laughing, and sounds from the rides. They can smell food from the food stalls.
2. Humans have senses to help inform us of our environment and keep us safe; students' own answers.
3. Students' own answers; animals also need air and (clean) water to survive.

Consolidation

Consolidation and reinforcement of the students' understanding of the topic can be undertaken using Topic quiz sheets B3, B4 and B5. This can be completed in class or as a homework task.

Topic quiz sheet answers

Sheet B3
1. Students' labelled drawings
2. Students' own answers

Sheet B4
1. goat – grass; bird of prey – mouse; butterfly – flower; frog – fly; leopard – antelope
2. food; salt
3. False; True; False

Sheet B5
1. sight – eyes, touch – hands, smell – nose, hearing – ears
2. George and Charlie, George, Charlie and Sam
3. look, listen, safe

Biology • Topic 2 Humans and other animals Student's Book answers

Student's Book answers

Pages 16–17 (2.1)
1 Students' own answers
2 Students' own answers
3 Answers might include: ears, eyes, noses, mouths, arms, legs, hands, fingers, etc.
4 They are moving, stretching, catching a ball, holding a rope, etc.

Pages 18–19 (2.2)
1 She is looking at and smelling the flowers. She is using her eyes and nose.
2 They are eating. They are using their sense of taste.
3 They are using their senses of touch, taste, sight and hearing.
4 Your sense of sight when you look at it. Your sense of touch when you peel it. Your senses of smell and taste when you eat it.
5 You cannot taste things as easily.

Pages 20–21 (2.3)
1 They are using their senses of touch, taste, sight and hearing. The family could also be using their sense of smell, i.e. they could smell the sea air, sand, etc.
2 They can see the beach, they can hear each other and the seagulls, they can taste the ice creams, they can feel the sand, sun and air around them. The family could also be using their sense of smell, i.e. they could smell the sea air, sand, etc.
3 They can hear traffic, talking, footsteps, etc. They can see people, cars, buildings, etc.
4 Your sense of sight lets you see the traffic. Your sense of hearing lets you listen for traffic or danger.
5 Yes.

Pages 22–23 (2.4)
1 Eye
2 Sight
3 Students' own answers; answers might include: watch TV, read a book, etc.
4 Students' own answers
5 Students' own answers

Pages 24–25 (2.5)
1 Goat, tiger, gorilla, human
2 Grass, meat, fruit, vegetables
3 Gorillas and humans
4 Answers might include: bottled water, tap water
5 No, because it is too salty.

Pages 26–27 (2.6)
1 Children, parents, grandparents, men, women, baby, girl, boy
2 They all have two eyes, two ears, a nose, a mouth, a body, etc.
3 They both have a mouth, a nose, a body, two eyes, two ears, etc.

Pages 28–29 (2.7)
1 They all have a mouth, a nose, a body, two eyes, two ears, etc.
2 They are different heights, they have different colour hair, skin, eyes. Two are boys, two are girls.
3 They grow longer and stronger.
4 Answers might include: walk, run, feed myself, dress myself, talk, read, etc.
5 Rebecca

Chemistry • Topic 3 Materials

Thinking and working scientifically

3.1 Similar or different?

Student's Book pages 32–33

Thinking and working scientifically
- *Carrying out scientific enquiry:* 1TWSc.01 Sort and group objects, materials and living things based on observations of the similarities and differences between them.

Resources
- Workbook page 24
- PCM C1: Sorting into groups
- selection of objects made from wood, metal, plastic and fabric
- sheets of paper

Classroom equipment
- two different books
- selection of toy bricks and/or marbles of different sizes and colours

Key words
- similar • different • feature • sort • group

Skills and connections

From an early age, children begin to sort objects, such as arranging dolls on a shelf or lining up toy cars on the floor. This desire to match similar items and organise them into a logical order seems to be a natural instinct. Although children will be unaware that they are doing so, they are already using their observations of similarities and differences and applying these to their everyday lives.

Students may have already done some work on sorting in mathematics; it will be useful to check with their mathematics teacher what they have done to see what foundation they have. Science lessons will build on this and will teach students to apply their skills of sorting and grouping to objects, materials and living things. Initially, this will involve simple features that students are familiar with, such as recognising similarities and differences in size, shape and colour. As their powers of observation, language skills and knowledge of the world develop, this will allow students to extend their sorting skills to more complex features such as type of material and the properties of a material.

In order to be able to sort items into groups, students will need to choose a feature to sort by. To do this, they need to observe and find the things in common. As well as being able to sort and group, students should be able to give sound reasons for their groupings. They should be encouraged to give as much detail as possible and to use science words where applicable. The skills of sorting and grouping are applied throughout the course. For example, in Topic 1 students sorted animals and plants, and living and non-living objects. In Topic 4 they will sort things based on sounds. This topic is about materials and provides ample practice in sorting and grouping.

Introduction

- Remind students of the work they have done in earlier units on finding similarities and differences between things, for example in Unit 1.2 (living and non-living) and Units 2.6 and 2.7 (human features). Write the key words *similar*, *different* and *feature* on the board and point out how each word is pronounced.
- Hold up two different books. Ask students: *What is similar about the books? What is different?* Elicit ideas and discuss as a class. Encourage as many varied ideas as possible and support higher-level critical thinking beyond simple features such as size, colour and shape, for example author, use of the book.
- Explain that in science, we often need to find similarities and differences in objects, materials and living things. We can then use these observations to compare things and *sort* them into *groups*, based on similar features. Add these two key words to the list on the board
- Remind students of the previous sorting work they have done, for example Unit 2.6 (sorting by height). Tell the class that they are going to learn more about sorting and grouping in this unit.

Teaching and learning activities

- Turn to Student's Book page 32. Work through the first two questions as a class. Ask: *What things are the same? What things are different?* Guide the students to establish that the toy bricks have similar and different features related to size, shape and colour. The activities in this

Chemistry • Topic 3 Materials

unit will help develop students' language skills as they learn to describe their observations. Provide students with suitable new vocabulary on the board for them to refer to.

- Ask students: *Do you sort any objects at home?* Guide students to realise that sorting can include things like arranging clothes in drawers, putting items in the correct place in the kitchen, tidying up toys into the correct place, etc. Discuss ideas and establish that objects are usually sorted based on similar features.
- Draw a large red circle, a large blue triangle, a small blue circle and a small red triangle on the board. Ask students: *How can you sort these shapes?* Elicit that they can be sorted in more than one way, depending on which feature you choose: shape or colour. Explain that when things have more than one similar feature, you need to choose which feature to sort by.
- Ask students to answer the next two questions on Student's Book page 32 in pairs. Discuss their ideas for sorting the objects, e.g. the marbles can be sorted based on size and colour but also features such as pattern or whether they are transparent or not. Give each group a selection of coloured toy bricks or marbles and allow them time to physically sort them into groups of their choosing. Circulate, reminding students to give reasons for their groupings.
- Tell the students to look at the pictures and answer the question on Student's Book page 33. Allow them time to discuss their answer in groups. Take feedback and check that the students have chosen to group by type of material. Some students may suggest alternative ways of sorting, such as properties or use. You can use this to reiterate that objects can be sorted in various ways using different criteria.

Graded activities

1 Give each pair of students a copy of PCM C1. Ask them to cut out the pictures and to sort them into two groups. As they do so, circulate and check that they are sorting the pictures correctly. This activity will also be a useful recap of content covered in Unit 1.3.

2 Ask students to look at Workbook page 24. Explain that it shows a Venn diagram with circles for glass and metal. Ask students to write the objects in the circle for the material that each object is made from. Some objects are made from both materials, for example a car has both glass and metal parts.

 Show the students a range of objects made from different materials. Allow them time to thoroughly explore the objects. They should handle the objects to feel them (are they heavy, light, rough, smooth?) and look at them in detail (what shape, size and colour are they?). Ask students to identify and describe each object in turn, taking ideas from the class until all appropriate adjectives have been used. Then ask: *How can you sort the things into groups?* Encourage the students to suggest as many suitable ways of grouping as possible.

Consolidate and review

- Ask the students to draw a Venn diagram on a large sheet of paper. Choose two materials. Call out names of objects and ask students to write the objects in the correct place in the diagram.

Differentiation

■ All students should be able to recognise the pictures and sort them into two groups based on obvious criteria.

● Most of the students be able to identify the material each object is made from, recognising that some are made from both materials, and then place them correctly in the Venn diagram.

▲ Some of the students should be able to accurately describe the objects in detail and suggest ways of sorting them. For those who need some prompting, circulate, asking questions to guide their thinking.

Chemistry • Topic 3 Materials

3.2 Properties of materials

Student's Book pages 34–35

Chemistry learning objective
- *Properties of materials:* 1Cp.01 Understand that all materials have a variety of properties.
- *Materials and their structure:* 1Cm.02 Understand the difference between an object and a material.

Thinking and working scientifically
- *Scientific enquiry: purpose and planning:* 1TWSp.02 Make predictions about what they think will happen.
- *Carrying out scientific enquiry:* 1TWSc.01 Sort and group objects, materials and living things based on observations of the similarities and differences between them; 1TWSc.02 Use given equipment appropriately; 1TWSc.03 Take measurements in non-standard units; 1TWSc.04 Follow instructions safely when doing practical work; 1TWSc.05 Collect and record observations and/or measurements by annotating images and completing simple tables.
- *Scientific enquiry: analysis, evaluation and conclusions:* 1TWSa.01 Describe what happened during an enquiry and if it matched their predictions.

Resources
- Workbook pages 25 and 26
- PCM C2: Describing materials
- Digital resource C1: Colours and textures

Classroom equipment
- wooden block
- selection of hard and soft materials
- selection of objects of different sizes and weights, and a simple means of weighing them
- paper bag, strong plastic bag
- plastic spoons, wooden spoons
- large sheets of paper, coloured pens or pencils
- shopping bags of similar sizes made from different materials: thin plastic, thicker plastic, paper, cloth
- loads with equal mass (e.g. marbles), plastic containers, small pieces of string, sticky tape, eye protection, large box

Key words
- material • object • properties

 Supervise students when they are handling breakable materials. Make sure the students use the strength test apparatus correctly and safely. Provide students with eye protection and put a large box on the floor to catch the weights. Make sure students use lengths of materials that end about 10 cm above the floor.

Scientific background

Materials are classified according to their *properties*: some are visible, some are determined by investigation or by using special equipment. The main visible properties are colour, transparency and texture. Less obvious properties include hardness, strength, flexibility and elasticity. The properties of a material determine its suitability for a particular purpose. Often several properties are taken into consideration, including cost and availability.

Introduction

- Use the topic opener photograph on Student's Book page 31 as a talking point. Ask the students: *What do you think this topic will be about?* Tell them that they are going to learn about different materials and their properties.
- Ask the class to look at the pictures on Student's Book page 34. Ask: *What do the different materials look like? What words would you use to describe them? In what ways are they different?* Encourage students to use a range of suitable adjectives to describe them. This activity is to assess the students' existing knowledge.
- Introduce and discuss the key word *properties* (singular: *property*). Show the class a material, for example a piece of wood. Ask: *What colour is it? What size is it? What does it look like?* Explain that colour is a property of a material. Write the word property on the board.

Chemistry • Topic 3 Materials

- Introduce more key words: *hard, soft, heavy, light, strong, weak, shiny, dull, rough, smooth*. Write them on the board and explain what each word means. Say we use these words to describe more properties of materials: the way they look and the way they feel.

Teaching and learning activities

- Ask: *What do we mean if we say* Point out some examples of *hard* materials around the classroom (e.g. wood, metal, plastic, glass). Give the students time to look at them. Then ask about *soft* materials (e.g. cotton, rubber, paper) and ask individuals for their ideas. Show the students two very different materials, for example a cotton t-shirt and a glass, and ask: *Which is harder?* Discuss the questions on Student's Book page 34 as a class. Establish why some objects need to be hard.
- *something is heavy or light?* Show the class a range of different objects. Ask a volunteer to put the objects in order, heavy to light, just by looking at them. Encourage the other students to comment. Ask: *Is there any pattern to the order?* It is likely that most of the big objects are placed as the heaviest and the small objects as the lightest. Check by weighing and move the objects so they are in the correct order of weight. Ask: *What is surprising about the order now?* Establish that heavy things do not always need to be big.
- Direct the students to Student's Book page 35 and discuss the picture. Ask: *What has happened?* (The bag was not strong enough to carry the shopping, so it has broken.) Give the students two bags, one made of paper and one made of fairly strong plastic. Ask: *Which bag is stronger? How do we know this?*
- Give each pair of students a set of cards showing different property words, cut from PCM C2. Show the students a range of objects made from different materials. Ask them to describe the objects using the words on the cards, such as *hard, soft, shiny, dull*.
- Make sure the students understand the question on Student's Book page 35. Ask them to discuss their answers in groups, then take feedback.

Graded activities

1 Give each pair of students some coloured pens and a large sheet of paper. Ask them to divide the paper into four equal sections. Tell them that they should make a poster to show four different objects: a hard object, a soft object, a smooth object and a rough object. They should choose an object that matches each property and draw a picture of it in that section of the poster. They can select from the objects they have seen in this lesson or any suitable alternative. Encourage them to add labels with each property name and the name of the object.

2 Ask the students to work in small groups. Give each group a plastic spoon and wooden spoon. Direct the students to Workbook page 25 and explain the activity. When they have completed the table, ask: *What have you found out?* Establish the properties the materials have in common and their differences.

3 Show the students some bags made from different materials and ask: *Which do you think will be the strongest?* Take a vote and put their predictions on the board. Hang a bag from a secured coat hook (or similar) and show the students what to do, starting with one weight in the bag and adding further weights one at a time. Say that they will be testing the different materials. Remind them that the test must be fair, so the weights must be the same size, and that the investigation ends when the material being tested is broken. They should record their results on Workbook page 26. Ask the students to use their results to rank the materials in order, from strongest to weakest. Ask: *Which was the strongest/weakest material? Was the test fair? Was your prediction correct?*

Consolidate and review

- Let the students complete Digital resource C1 to consolidate their learning.

Differentiation

■ All of the students should be able to identify an object for each of the four properties with little prompting.

● Most of the students should be able to work together, asking and answering questions and using first-hand observations to compare the two materials.

▲ Some of the students should be able to make predictions about the outcome of the investigation and then follow instructions to find out if they were correct.

Chemistry • Topic 3 Materials

3.3 More properties

Student's Book pages 36–37

Chemistry learning objective

- *Properties of materials:* 1Cp.01 Understand that all materials have a variety of properties.

Thinking and working scientifically

- *Scientific enquiry: purpose and planning:* 1TWSp.01 Ask questions about the world around us and talk about how to find answers.

Resources

- Workbook pages 27, 28 and 29
- PCM C3: Cardboard glasses

Classroom equipment

- glass object
- cup of water, selection of absorbent and less absorbent materials, such as sponges, dishcloths, paper towels, tissues, nylon, cotton, wool, polyester
- selection of objects made from different materials
- selection of different transparent plastics
- sticky tape
- large sheets of paper

Key words

- transparent • waterproof • stretch • elastic • bend
- flexible • absorbent

 Supervise students when they are handling breakable materials such as glass. They should not handle any objects with sharp edges or points.

Scientific background

The property of transparency (see-through) refers to letting light through. This means that the light rays pass through the material without being changed in any way. Glass, some plastics, air and water have this property. *Transparent* materials can be coloured. Some materials are *waterproof*, meaning they do not allow water to penetrate or pass through them. Some materials are not waterproof but are coated with a substance that is, which makes them waterproof. Substances such as rubber, wax and polyurethane are such coating materials. Materials that are not waterproof absorb water; *absorbent* materials soak up water and become wet. Absorbent materials are useful for a range of purposes, such as clearing up spillages and keeping things moist. Some materials are stretchy or *elastic*: they can be *stretched* but will return to the original shape once the force has been removed. The fabric that is called 'elastic' is not the only material that has this property; many other materials have elastic properties and they are used for many things, from clothes to car tyres. Some materials can be squashed or *bent* while others cannot. A material that is very squashy and bendy is *flexible*.

Introduction

- Review the key word from the previous lesson, *properties* (singular: *property*). Show the class an object made of glass. Ask: *What properties does this object have? What words would you use to describe it?* Write the students' suggestions on the board. Ask questions to elicit that we can see through glass. Tell students that the property word for this feature is *transparent*. Tell them that they are now going to learn about some more properties of different materials.

- Introduce some more key words for this unit and explain what each word means. Write them on the board and point out the pronunciation and spelling of each one.

Teaching and learning activities

- Ask the students to look at the picture on Student's Book page 36. As a class, discuss the transparent objects in the picture. Although the word *glass* is not introduced until Unit 3.4, most students will be familiar with this material from everyday life. Ask the students to think about how glass is used in the world around them. Let them name as many different objects made from glass as they can and make a class list on the board. Ask the students to talk with their partner about what makes glass a useful material. Can they name some other properties of glass such as hard and smooth?

- Ask: *What does waterproof mean?* Explain that some materials (e.g. rubber) are waterproof and do not let water through them. Ask: *What can we do to test if a material is waterproof?* (Pour water

Chemistry • Topic 3 Materials

on to it.) *What questions can we ask?* (For example: *Does it let water through?*)

- Spill a small cup of water on the floor, in front of the class. Ask the students if they have ever spilled anything. What did they do? What can they suggest to use in order to clear it up. Use some of their ideas and try out a range of different materials. Then ask students to talk with their partner to answer the questions: *Why was the paper towel (or sponge or tissue) useful for clearing up the water? Was the paper towel waterproof? Why do you think this?* Discuss the word *absorbent* and clarify its meaning, using the demonstration as an example to show that it means able to take in water. Then ask the students to answer the questions on Student's Book page 36.

- Establish that the term *flexible* refers to a material that can *bend* easily without breaking and that an elastic material returns to its original shape after it has been stretched. Ask the students to think of as many flexible and elastic materials as they can. List them on the board. Ask the students *What questions can we ask to find out more about flexible materials?* (For example: *Are flexible materials strong or weak? Are elastic materials hard or soft? Are flexible materials also elastic?*) Say: *Talk to your partner about what you could do to find the answers to the questions.* (For example, students might suggest trying to mark or scratch the material with a nail to see if it is hard or soft. They might suggest trying to bend the material using their hands to see if it goes back to its original shape.)

- Make sure the students understand the questions on Student's Book page 37. Allow them to discuss their answers in groups, then take feedback.

Graded activities

1 💬 Ask the students to complete the activity on Workbook page 27. They should look at the different objects and decide which property word(s) best describe each one. Circulate to check that they are matching the correct words to each object. Ask questions to guide their thinking. Once they have completed the activity, you can ask them to explain their reasoning.

2 Show the students a selection of absorbent materials, such as sponges, paper towels, hand towels and mops, etc. Tell them to choose three of the materials. They should then draw and label a picture of each one on Workbook page 28. Ask students if they can name any more absorbent materials and say what each material can be used for. Recap that all absorbent materials are used for soaking up water or other liquids. Ask: *What do you think happens to the water? What property word do we use for a material that can do this?*

3 Provide the students with a range of objects made from different materials and a large sheet of paper. Allow them time to explore the objects. Then ask them to choose three objects from the selection. Tell them that they need to draw each one on page 29 of their Workbook and describe how it looks and feels, and what its properties are. Encourage the students to use as many descriptive words as possible. Point out that they can use the key words on the board for ideas.

Consolidate and review

- Help the students make some glasses out of transparent plastic. Have some cardboard frames already cut from PCM C3. The students use tape to stick the different plastics onto the frames. Allow them to vote on which glasses are the best.

Differentiation

■ All of the students should be able to correctly match the property words to each object.

● Most of the students should be able to draw a reasonable representation of each material and label it correctly.

▲ Some of the students should be able to accurately describe their chosen objects and identify the properties. For those who need some prompting, circulate, asking questions to guide their thinking.

Chemistry • Topic 3 Materials

3.4 What material is it?

Student's Book pages 38–39

Chemistry learning objective

- *Materials and their structure:* 1Cm.01 Identify, name, describe, sort and group common materials, including wood, plastic, metal, glass, rock, paper and fabric; 1Cm.02 Understand the difference between an object and a material.
- *Properties of materials:* 1Cp.02 Describe common materials in terms of their properties.

Thinking and working scientifically

- *Scientific enquiry: purpose and planning:* 1TWSp.01 Ask questions about the world around us and talk about how to find answers.

Resources

- Workbook pages 30 and 31
- Slideshow C1: Wood
- Video C1: Glass shaping

Classroom equipment

- samples of different materials: as many as possible from wood, stone, paper, cardboard, fabric, wool, cotton, metal, plastic, glass
- range of objects made from glass, such as beaker, thermometer, drinking glass and bottle
- selection of different objects made from wood
- selection of objects made from wood, metal, fabric and plastic

Key words

- material • object

 Supervise students carefully when they are handling breakable materials such as glass. They should not handle any objects with sharp edges or points.

Scientific background

There are many different types of materials. This unit will introduce the students to some of the more common materials, such as *wood, rock, paper, fabric, metal, plastic* and *glass*. *Wood* comes from trees and is very useful. *Paper* and *cardboard* are made from wood. *Stone* is quarried from rock faces. Stone (or rock) is very hard and strong so is often used for making things such as building blocks and paving stones. *Fabric* is the generic term for types of cloth or textile. Different types of fabric made from fibres originating from plants and animals include *cotton* (from cotton plant) and *wool*. Wool is the generic term for fibres obtained from sheep and certain other animals. There are also manufactured fibres, such as nylon and polyester, which the students will look at in more detail in Unit 3.7. *Metals* are materials that come from rocks called metal ores. Metals are very hard, strong and shiny and can be made into many different things. There are many different types of metals, some with unique properties. *Plastics* are made from crude oil. Plastics are chemically manufactured to have different properties and can be shaped and moulded, making them suitable for a very wide range of uses. *Glass* is an unusual material. It is made from sand, sodium carbonate and limestone, heated to a high temperature and cooled. Most glass is transparent so it is usually used to make windows.

At this stage, the students will learn the names of these materials and relate them to properties they learned in previous units; they do not need to understand how the materials are sourced or made.

Introduction

- Introduce the key words *material* and *object*. Then introduce some examples of different materials: *wood, stone, paper, cardboard, fabric, wool, cotton, metal, plastic, glass*. Write them on the board and point out how each word is spelled and pronounced. Explain that there are many different types of materials in the world and that these are the names of a few common materials.
- Point to objects in the classroom. Ask the class to identify each object and the material it is made from, for example, a window made from glass. Stress that wood, cotton, metal, etc., are all *materials*. Emphasise the difference between what an object is (and is used for) and the material it is made from. For example, say: *This is a spoon. It is used for eating. It is made from metal.* Have some samples of materials that have not been made into everyday objects, for example, pieces of plastic, wood, metal, to help identify the material and distinguish it from the object.

Chemistry • Topic 3 Materials

Teaching and learning activities

- Ask the students to look at the pictures on Student's Book page 38. Ask what they can see, what each object is and what they think it is made from. Discuss their ideas as a class. Allow the students time to answer the questions in their groups. Take feedback and elicit the objects made from wood and plastic, and the different uses of fabric. Ask the class: *Why are books made from paper?* Accept any reasonable suggestions, such as: because you can print on paper; paper is easy and cheap to make; paper can be cut to a particular size. Ask: *Why are books not made from glass?* Again accept any reasonable answers.
- Show the students some objects made from wood. Ask the class where wood comes from. (trees) Show the class Slideshow C1, of trees being felled and some wooden products. Ask the students to think of some ways in which we use wood in the world around us. Ask: *Why are chairs made from wood? What would happen if a chair was made from paper?* (it would collapse) *How do you know this?* (we can test it; we know paper is not as strong as wood, etc.). Establish that wood can be shaped into many different objects. Point to a chair and say: *Even though it has been made into an object, it is still wood; it came from a tree.* Explain that wood is also used to make cardboard and paper.
- Show the students some objects made from glass. Include a beaker, a glass thermometer, a drinking glass and a bottle. Explain that a lot of glass is blown into shape, either by mouth or by machine. Show Video C1, of glass shaping and how it can be formed into different shapes. Ask: *What did the man make from glass?* Ask the students to think of as many different objects made from glass in the world around them as they can and to say what the properties of glass are. Establish that glass is *transparent*, which means they can see through it.
- Show the students the range of materials. Ask if they can think of any objects made from plastic at home or in school. Ask: *What properties do the different types of plastic have?*
- Make sure the students understand the questions on Student's Book page 39. Ask them to discuss their answers in groups, then take feedback.

Graded activities

1 Ask the students to complete the task on Workbook page 30. Ask them to tick the material that would be best for making each object. Remind them to think about the most important property for the object to do its job and to assess which material offers that property.

2 Invite the students to go on a materials hunt around the classroom, looking for things made from wood, metal, fabric and plastic. Before the lesson, hide some materials around the classroom for the students to find. The students should look at the pictures on Workbook page 31 and then search for similar items. Ask: *What did you find that is made from this material? Where did you find it? What is the object?*

3 Tell the students that a toy company is making a new car for 2-year-old children. They want to know which materials they should make it from. Ask: *What materials should they use? Why?* Ask the students to design a toy with their partner, using only wood, metal and plastic. They should say how they would use the materials on these toys. Ask: *Why do you think the materials you have chosen would be the best?* Write a list of property words on the board to help them, such as *hard, soft, strong, rough, smooth, light, flexible, transparent*.

Consolidate and review

- Tell the students you are going to say the names of some objects. Some of them will be made from plastic and some will not. They should stand up when they hear one that can be made from plastic.

Differentiation

■ All of the students should be able to decide which material each object is made from.

● Most of the students should be able to locate some objects made from the various different materials and sort them into the correct category.

▲ Some of the students should be able to work collaboratively to share ideas and think creatively. They will be able to choose appropriate materials for the different parts of their toy and then provide reasons, demonstrating that they have understood the different properties of each material.

Chemistry • Topic 3 Materials

3.5 More materials

Student's Book pages 40–41

Chemistry learning objectives
- *Materials and their structure:* 1Cm.01 Identify, name, describe, sort and group common materials, including wood, plastic, metal, glass, rock, paper and fabric.
- *Properties of materials:* 1Cp.01 Understand that all materials have a variety of properties; 1Cp.02 Describe common materials in terms of their properties.

Thinking and working scientifically
- *Scientific enquiry: purpose and planning:* 1TWSp.02 Make predictions about what they think will happen.
- *Carrying out scientific enquiry:* 1TWSc.05 Collect and record observations and/or measurements by annotating images and completing simple tables.

Resources
- Workbook page 32
- Video C2: Making concrete
- PCM C4: Concrete or glass?
- Digital resource C2: Metal, wood, plastic

Classroom equipment
- cards, large sheets of paper, coloured pens
- selection of objects made from wood, metal, stone, fabric, leather and rubber
- pictures of different bridges
- materials for making model bridges, such as drinking straws, cocktail sticks, rope or string, sheets of paper, modelling clay, corrugated card, etc.
- toy car
- scissors, glue, old magazines

Key word
- material

 Supervise the students when they use scissors and glue. If you take the students on a walk around the school grounds, ensure they are safe and that they stay together.

Scientific background

Concrete is a human-made (manufactured) material. *Clay* is a type of heavy, sticky soil that becomes hard when it is baked. It is often used to make *brick*. Natural *rubber* is made from a liquid called latex that comes from certain plants. Manufacturing processes are used to turn latex into a much more versatile material. About 30 per cent of the rubber used today is natural rubber. The other 70 per cent comes from crude oil, from which the rubber is made synthetically. *Leather* is a durable and flexible material, created by tanning animal hide or skin, commonly from cattle.

Sometimes materials may be combined to produce a new material that has different properties. For example, copper is a soft metal and zinc is rather brittle, but together they make the alloy brass, which is hard and tough. Manufacturing often involves a change that leads to a new set of desired characteristic properties. An example of a modern, manufactured material is that used for airbags in the safety systems of many cars. In the event of an accident, these inflate automatically to protect the driver and passengers. Airbags must be made of very flexible material, so they can blow up quickly. They also need to be strong, so they do not burst under impact.

Introduction
- Ask the students to discuss what a material is and to name some different materials from the previous lesson. It is important to ensure that the students have understood that *material* is not just another word for fabric or cloth. Make a list on the board of common materials from the previous lesson.
- Remind the students what wood, metal, stone, rock and fabric look like. Show them an object made from each material and ask: *What is this object made from?* Then introduce the words *concrete*, *brick*, *clay*, *rubber* and *leather*, and explain that they are the names of some common materials. Write the words on the board and point out how each word is spelled and pronounced.
- Ask the students to select one of the materials from the board and write its name on a piece of card. Describe your journey to school and ask

Chemistry • Topic 3 Materials

them to hold up their card every time an object is mentioned that is made from their material.

Teaching and learning activities

- Show the students objects that are made from rubber and leather, and ask them to describe the differences between the two materials. Ask the students to look at the first two pictures on Student's Book page 40. Ask: *Can you think of any other objects made from rubber and leather?*
- Show the class Video C2 about how concrete is made. Explain that once the concrete is made, it can be moulded into different shapes, such as blocks or paving stones, and left to set into shape. Discuss some uses of concrete with the students.
- Tell the students they are going to go on a tour of the school grounds. Let them first make some predictions about the concrete or glass objects they might see. Ask them to record their predictions on PCM C4. On the tour, students should add to their lists any other objects that they did not predict. Ask: *Which of your predictions were correct? Which were not correct? Why was this? Did you find any objects that you did not expect? Which material was more common, concrete or glass? Why do you think this was?*
- Ask the students to look at the pictures on Student's Book pages 40–41. Talk about the modern materials used to make the tablet, the helmet and the food box. Try to identify some of the materials used. Discuss the properties of the materials. Ask the students if any other materials could have been used instead.

Graded activities

1 Give each student a sheet of paper and some coloured pens. Ask them to imagine a future world with no wood or wood products. Ask them to draw a picture of what their bedroom would look like. Circulate to make sure they are not drawing anything that is made from wood. Ask questions such as: *If we cannot use wood to make a bed frame, what could we use instead?*

2 Ask the students to think of three objects at home that are made from different materials. Explain that the objects can be made from a single material or several materials, but encourage them to choose at least one object that is made from multiple materials. Ask students to draw and label each object on page 32 of their Workbook, and to describe the properties of the materials used.

3 Show the students some pictures of different bridges. Ask the students to identify the materials used for each of them. Establish that the main materials for building bridges are metal, concrete and wood. Tell the students that they are going to try making their own bridges out of different materials. The bridge should be about 20 cm long and 10 cm wide. Show them these dimensions. Give each group some materials. Ask them to work together to design a bridge that will stand up and support the weight of a toy car going over it. Ask the students to present their bridges to the class. Ask the class to predict which will be the strongest bridge.

Consolidate and review

- Arrange the students into groups. Give them two large sheets of paper, scissors, glue and some old magazines. Ask them to put the title *Metal* on one sheet of paper and *Concrete* on the other. Each group then cuts out pictures of objects made from metal or concrete and use them to create a collage or display of materials. Ask the students to describe the objects and say what they are used for.
- Let the students complete Digital resource C2 to consolidate their learning.

Differentiation

■ All of the students should be able to think creatively and relate this to what they have learned about wood and its various uses. Students should also be able to use their knowledge of properties of other materials to suggest suitable replacements for wooden objects, such as a metal bed frame.

● Most of the students should be able to identify which materials each of the three objects are made from and correctly describe their properties. Some may need a little prompting if they cannot recall all the properties they have learned in this topic.

▲ Some of the students should be able to work collaboratively to share ideas and think creatively about which materials to use in their bridge design. As long as students give sensible reasoning for their choice of materials, there are no right or wrong answers and all designs should be encouraged.

Chemistry • Topic 3 Materials

3.6 Sorting materials

Student's Book pages 42–43

Chemistry learning objective
- *Materials and their structure*: 1Cm.01 Identify, name, describe, sort and group common materials, including wood, plastic, metal, glass, rock, paper and fabric.

Thinking and working scientifically
- *Carrying out scientific enquiry*: 1TWSc.01 Sort and group objects, materials and living things based on observations of the similarities and differences between them.

Resources
- Workbook pages 33–34 and 35
- PCM C2: Describing materials
- PCM C5: Sorting materials

Classroom equipment
- selection of objects made from wood, metal, stone, plastic, glass and fabric
- drinking glass, glass vase
- old magazines, scissors, glue
- wooden bricks or blocks of different sizes, shapes and colours

Key words
- sort • group • properties

 Supervise the students when they use scissors and glue. Supervise them carefully when they are handling breakable materials such as glass. They should not handle any objects with sharp edges or points.

Scientific background

Materials are classified according to their *properties*. There are many different properties: some are visible, some are determined by investigation or by using special equipment. The main visible properties are colour, transparency and texture. Less obvious properties include strength, hardness, flexibility and elasticity. We can group materials by their type, such as metal or wood, or by a property that they have in common, such as being transparent or smooth. It is sometimes possible to *sort* a material into more than one *group*. This is because a material can have more than one property. For example, wool is both soft and flexible.

Introduction

- Show the students a range of objects from previous lessons. Call up different students and ask them to take it in turns to select a plastic object. When all the plastic objects have been chosen, hold one up and ask: *What made you decide this was a plastic object? What properties does it have?* Arrange all the plastic objects in a group together. Explain that we can sort objects into groups.
- Introduce the key words: *sort, group, properties*. Write them on the board and point out how each word is spelled and pronounced. Ask the students to repeat the words after you. Tell the students to look at the group of plastic objects. Ask: *Are all these objects made from plastic?* (yes) Establish that we can group objects based on their material.

Teaching and learning activities

- Ask students to look again at the group of plastic objects. Add a drinking glass and a glass vase. Ask: *Are all these objects made from plastic?* (no) Then ask: *Are all these objects transparent?* (Ensure they are.) Establish that we can also group according to different properties. Explain that it is possible for different materials to be grouped together if they share the same property.
- Ask the class to look at the picture on Student's Book page 42. Ask: *What can you see in the picture?* Talk about the different materials and objects in the picture, and answer the questions as a class.
- Using the range of objects from the introduction, call up different students and ask them to take it in turns to pick out all the wooden objects. When all the wooden objects have been chosen, group them together on the table. Then ask the students to pick out all the metal objects, and do the same again. Use this activity to reinforce the

Chemistry • Topic 3 Materials

concept that objects can be sorted in groups by type of material.

- Give each group a tray with a selection of objects made from different materials, including wood, metal, plastic, fabric, glass and stone (rock). Let them examine the objects and use descriptive words to explain the properties. Take feedback and compile a class list on the board.
- Ask the students to group the objects according to some common properties, for example, hard, soft, etc. In each case, ask them to justify their groupings and which objects they have chosen. Ask: *Why have you grouped those object together? Is everything in the group made from the same material?* Ask the students to find other ways of grouping the same objects, by using different properties. Use this activity to reinforce the concept that materials can be sorted into groups by property.
- When the students have grouped their materials, ask them to look closely at one of their groups by asking: *Are all the samples of wood the same? Are all the pieces of plastic the same colour?* It is important that the students begin to see that there are different types of wood, plastic, etc. (They will learn more about this in Unit 3.7.) Emphasise the difference between the object itself and the material it is made from.
- Give each group a set of word cards cut from PCM C2, of different properties. Place wooden, metal, cotton and plastic objects in front of the students. Ask them to place the correct property word cards with each material; some can be used more than once. Encourage them to talk about their choices and to give some examples.
- Make sure the students understand the questions on Student's Book page 43. Let them discuss their answers in groups.

Graded activities

1 Give each group of students a selection of old magazines, some scissors and glue. Ask them to look for some pictures of objects made from wood, metal, plastic and fabric. They should then cut out the pictures and sort them into groups before sticking them in the correct places on Workbook pages 33–34. Remind students that *fabric* covers a wide range of materials, including wool and cotton.

2 Provide the students with a selection of different objects to sort. They should work together to place all of the wood, plastic, metal, stone and fabric objects into separate piles of the same materials. Ask: *What features of the materials made you sort them into these groups?* Talk about the differences in the materials that help to sort the objects. Ask: *What about the colour of the material? What did the material feel like? Was it a strong material?* Recap the different properties that these materials have. Ask if the students can find more than one way to sort the objects. If necessary, remind them that we can sort by type of material and by property.

3 Ask the students to complete the activity on Workbook page 35. Explain that they need to decide which item is the odd one out from three given objects or materials. This task is made harder because they also have to give a reason for their ideas. Ask questions to guide their thinking: *What do two of the objects have in common? What are they made from? Where does the material come from? Is it made from metal? What do the objects look like?*

Consolidate and review

- Give pairs of students a copy of cards cut from PCM C5. Ask them to sort the pictures into fabric, metal, wood and stone.

Differentiation

■ All of the students should be able to identify pictures of objects made from the four different materials and correctly sort them into groups, with little prompting.

● Most of the students should be able to explore the selection of objects and use their knowledge of materials and their properties to sort them into appropriate groups.

▲ Some of the students should be able to think critically to choose which is the odd one out and provide sound reasoning for their choice. Some students may struggle to do this, especially as there is a mix of materials and objects, so circulate and offer support as necessary.

Chemistry • Topic 3 Materials

3.7 Making smaller groups

Student's Book pages 44–45
Chemistry learning objective
- *Materials and their structure:* 1Cm.01 Identify, name, describe, sort and group common materials, including wood, plastic, metal, glass, rock, paper and fabric.

Thinking and working scientifically
- *Scientific enquiry: purpose and planning:* 1TWSp.02 Make predictions about what they think will happen.
- *Carrying out scientific enquiry:* 1TWSc.01 Sort and group objects, materials and living things based on observations of the similarities and differences between them; 1TWSc.04 Follow instructions safely when doing practical work; 1TWSc.05 Collect and record observations and/or measurements by annotating images and completing simple tables.
- *Scientific enquiry: analysis, evaluation and conclusions:* 1TWSa.01 Describe what happened during an enquiry and if it matched their predictions.

Resources
- Workbook page 36–37 and 38
- Slideshow C2: Traditional clothing

Classroom equipment
- samples of wool, cotton, silk, polyester, nylon
- woollen scarf
- selection of plastic objects, e.g. plastic bags, plastic bottles, soft plastic film, hard plastic buckets
- selection of different types of paper, e.g. newspaper, writing paper, poster paper, wrapping paper, cardboard, tissue paper, greaseproof paper
- small samples of plastic cut from food wrap, plastic bags and plastic bottles
- different types of paper for collages, scissors, glue

Key words
- fabric • plastic

 Supervise the students when they use scissors and glue. Ensure there are no sharp edges on the cut plastic bottle samples.

Scientific background

There are different types of some materials. Each type has different properties, so we can sort the different types into smaller groups. For example, wool, cotton, silk and nylon are all types of fabric. Different processes are used to make fabric with different properties, giving them a variety of uses.

Plastics are very useful materials because so many different types of plastics can be made, all with particular properties. Some are waterproof, some are heat-resistant and some can withstand corrosion from different chemicals, such as acids. Most are light, long-lasting and hardwearing, making them suitable alternatives to metal.

There is a huge variety of different uses for paper, owing to the different properties available. Different chemicals are added to provide different textures and qualities. Examples of paper include cardboard, writing paper, paper bags, tissues and so on.

At this stage, students do not need to understand how different types of different materials are made; they simply need to appreciate that the different types exist and have different properties, making them suitable for a range of uses.

Introduction

- Introduce some more key words: *wool, cotton, polyester, silk, nylon*. Write them on the board and point out how each word is spelled and pronounced. Ask the students to repeat the words after you. Explain that there are different types of some materials and these are all types of *fabric*. Then ask the students to look at the pictures on Student's Book page 44. Ask: *Can you recognise any of the types of fabric in the pictures?* Answer the questions as a class.
- Show students samples of wool, cotton, silk, polyester and nylon. Ask students to feel them and to compare the similarities and differences. Encourage them to use property words such as *soft, flexible, smooth*, etc. Tell the class that they are now going to learn about different types of fabric, plastic and paper.

Chemistry • Topic 3 Materials

Teaching and learning activities

- Ask the students: *Where does wool come from?* Show them something made from wool. Discuss with the students the fact that, even though it has been made into an object, it is still wool. Explain that wool often comes from sheep.
- Ask the students to look at the pictures at the bottom of Student's Book page 44. Ask: *What do you know about plastic?* Take answers and discuss some of the properties of different types of plastics (flexible, soft, hard, transparent, etc.). Ask: *Why is plastic so useful? Name some of the many different ways we use plastic.* Make sure the students understand that there are many different types of plastic, each with different properties and uses. Illustrate this with a selection of different plastic objects. Explain that these objects are all made from plastic, but the plastic has different properties. Allow the students time to explore the objects. Ask: *In what ways are they similar? In what ways are they different? What do we use each one for?* Ask them to sort the different types of plastic into groups based on their own choice of criteria.
- Show the class a range of different types of paper. Invite the students to feel the paper samples and describe their properties. Ask: *Where does paper come from?* (trees) *Why do we need so many different sorts of paper?* (It has many different properties and therefore uses.)
- Direct the students to Student's Book page 45. Ask: *What is the girl in the picture doing? What type of paper do you think she is using?*

Graded activities

1 Show Slideshow C2, of different traditional clothing and costumes from around the world. Prompt students to describe the colours and textures of the different clothes the people are wearing. Ask: *Can you think of any other uses of fabric?* Write a class list on the board.

2 Provide students with a selection of paper to explore and ask them to choose six different samples. They should write the properties of each one on Workbook page 36 and try to name the specific types of paper. (Provide English translations as necessary.) Then the students should say which paper they think is the strongest, thinnest, smoothest, heaviest, shiniest and most flexible. Finally, ask them to glue their six different types of paper on Workbook page 37.

3 Provide the students with small samples of food wrap, plastic bags and pieces cut from plastic bottles. Ask the students to work together to test each piece of plastic by a) trying to look through it, b) stretching it, and c) trying to break it. They record their results on Workbook page 38. Take feedback and discuss their findings.

Consolidate and review

- Ask students to work in small groups to make a poster about an environment of their choice using different types of paper. Ask: *What types of paper will you use and why?* For example, they could use shiny paper for metals, blue tissue paper for water and corrugated cardboard for wood. Once they have agreed on a plan, ask them to make their poster.

Differentiation

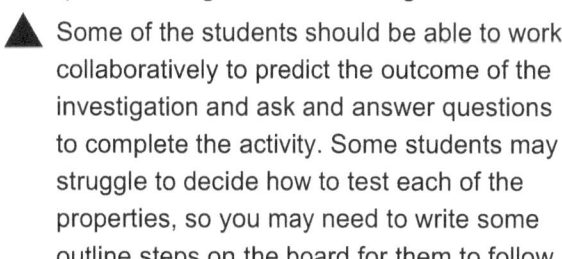

All of the students should be able to use their knowledge of the properties of different types of fabric to suggest some different uses of fabric.

Most of the students should be able to describe and compare the properties of the different types of paper. Circulate, making sure they are doing so correctly and asking questions to guide their thinking.

Some of the students should be able to work collaboratively to predict the outcome of the investigation and ask and answer questions to complete the activity. Some students may struggle to decide how to test each of the properties, so you may need to write some outline steps on the board for them to follow.

Chemistry • Topic 3 Materials

3.8 Materials can change shape

Student's Book pages 46–47
Chemistry learning objective

- *Changes to materials*: 1Cc.01 Describe how materials can be changed by physical action, e.g. stretching, compressing, bending and twisting.

Thinking and working scientifically

- *Scientific enquiry: purpose and planning*: 1TWSp.01 Ask questions about the world around us and talk about how to find answers; 1TWSp.02 Make predictions about what they think will happen.
- *Carrying out scientific enquiry*: 1TWSc.04 Follow instructions safely when doing practical work; 1TWSc.05 Collect and record observations and/or measurements by annotating images and completing simple tables.
- *Scientific enquiry: analysis, evaluation and conclusions*: 1TWSa.01 Describe what happened during an enquiry and if it matched their predictions.

Resources

- Workbook pages 39 and 40
- Video C3: Bread dough
- Video C4: Airbags
- PCM C6: Elastic bands

Classroom equipment

- modelling clay (or similar)
- wooden blocks
- a selection of everyday materials for students to test, such as a plastic comb, wooden pencil, elastic band, paper, etc.
- elastic bands and paper
- paper and pencils
- a selection of elastic bands (small, medium, long, narrow, wide, different colours)

Key words

• material • squash • bend • twist • stretch

 Supervise students carefully if they are working with elastic bands. Ensure they do not overstretch the bands and cause them to break. When testing elastic materials, they should wear eye protection.

Scientific background

The way we can change materials is largely dependent on their properties. The shape of certain materials can be changed by applying a force to the material. Materials are pushed together when you squash or squeeze (compress) them, and pulled apart when you stretch them. Materials that are flexible or malleable can be twisted and bent easily. Harder materials such as metals can be shaped by force if they are heated first.

Introduction

- Show the class Video C3 of a baker working dough. Ask what the baker is doing to the dough and elicit the answers.
- Use Video C4 to show the students that materials (the airbag in this case) can be stretched to make a round shape and also squashed (when the dummy hits the airbag). Discuss why it is important for this material to be able to change shape.
- Use the photographs on Student's Book pages 46–47 to revise the names of some common materials. Before you answer the questions as a class, ask students to predict which of these materials can change shape. Ensure they are correctly distinguishing between the objects and the materials they are made from.

Teaching and learning activities

- Ask the students to show (with their hands) what the actions squash, bend, twist and stretch mean. Ask questions such as: *What would happen to a paper cup if we squashed it? What would it look like?* Elicit that the act of squashing would change the shape of the cup. Explain that the other actions can also change the shape of some materials.
- Let students investigate how to change the shape of materials by modelling some clay. First, ask each student to make a long pencil shape. They should lay it on the table, hold one end still and pull the other end. Ask: *What happens?* (It stretches.) *Now pull both ends gently with both hands. What happens? What happens if you pull a bit harder? And harder again?* (It breaks.) *Does the same thing happen with a real pencil?* (No.)
- Ask: *What does 'squash' mean? Can you show me with your modelling clay?* Establish that squashing

Chemistry • Topic 3 Materials

an object means squeezing it in many directions. Ask each student to squash a ball-shaped piece of modelling clay gently, then harder until it is flat. Ask: *Does the same thing happen with a real ball?* (No.) Say: *Some balls can be squashed but others cannot. What is the difference?* (The action only works on materials that are flexible. Hard materials will not change shape.)

- Ask the students to demonstrate bending and twisting with the clay. Ask questions to make sure that the students are confident with all four action words they have learned and that they can use them correctly. Write the key words on the board. Point out how each word is spelled and practise pronunciation.

- Give each group of students a lump of modelling clay. Ask them to complete the activity on Workbook page 39, exploring the actions that can change the shape of the material. Students should then consider whether they can change the shape of a wooden block in the same way as they changed the shape of the clay. You could give each group a wooden block to try the actions on. Ask questions to guide their thinking: *What property does the clay have that allows you to squash it? Does wood also have this property? What are the properties of wood?*

Graded activities

1 Give each pair of students an elastic band and a piece of paper. Ask them to work together to answer the questions orally. They should try stretching and squashing both materials and observe the differences. The elastic band returns to its original shape because it is elastic; the paper does not do this because it does not have this property. Invite pairs to feedback to the class and discuss their findings.

2 Give the students a set of everyday materials: include some that will change shape and some that will not change shape when human force is applied. Ask the students to work in groups to complete the activity on Workbook page 40. Make sure they spend some time answering the first two questions before they do the practical investigation. Invite some groups to share their findings with the class and address any misconceptions. Ask: *What did you learn about the materials?*

3 Give each pair a selection of elastic bands to investigate. Ask them which band they think will stretch the most and why. Tell them to write their prediction and reason on PCM C6. They should then test their predictions. To do this, they should take turns to put each band over two pencils. As one student holds one pencil with its point at the edge of a sheet of paper, the other uses the second pencil to extend the band to its full unstretched length and makes a mark on the paper. The first student keeps their pencil still and the second uses their pencil to stretch the band as far as they can, and makes a second mark. They should describe each elastic band and record their findings in the table on PCM C6. Students will need to be able to measure with a ruler. Ask: *Does width, length or colour affect how much the bands stretch? Which elastic band stretched the most?* Tell the class that they will learn more about stretching in Unit 3.10.

Consolidate and review

- Ask students to use a small cushion to demonstrate all the ways in which its shape can be changed. All the actions can be applied to a cushion to varying degrees: squashing, bending, twisting and stretching.

Differentiation

■ All of the students should be able to explore the two different materials by stretching and squashing them and observing their different behaviour.

● Some of the students should know the meanings of the action words and be able to test materials by 'doing' the actions. Some students may not realise that more than one action can be applied to the same material. If this is the case, ask questions to help them realise this.

▲ All of the students should be able to give a prediction, but not all of them may arrive at the correct answers. The aim of the activity is for students to realise that you can test materials to see how much they stretch, and that you need to measure and record the results in order to say which material stretches the most.

Chemistry • Topic 3 Materials

3.9 Squashing and bending

Student's Book pages 48–49
Chemistry learning objective
- *Changes to materials*: 1Cc.01 Describe how materials can be changed by physical action, e.g. stretching, compressing, bending and twisting.

Thinking and working scientifically
- *Scientific enquiry: purpose and planning:* 1TWSp.02 Make predictions about what they think will happen.
- *Carrying out scientific enquiry:* 1TWSc.04 Follow instructions safely when doing practical work; 1TWSc.05 Collect and record observations and/or measurements by annotating images and completing simple tables.
- *Scientific enquiry: analysis, evaluation and conclusions:* 1TWSa.01 Describe what happened during an enquiry and if it matched their predictions.

Resources
- Workbook pages 41 and 42
- Video C5: Shaping metals
- Slideshow C3: Bridge shapes

Classroom equipment
- modelling clay
- soft ball or bathroom sponge
- a selection of objects, some that can be squashed by varying amounts and others than cannot
- large sheets of paper, coloured pens or pencils
- cricket balls and tennis balls
- old magazines, scissors, glue

Key words
- squash • bend • flexible

 Supervise the students when they are handling materials. They should not handle any objects with sharp edges or points. Supervise the students when they use scissors and glue.

Scientific background

Certain solids can be *squashed* (or compressed) by an applied force. Some of these objects will remain compressed when you stop applying the force. For example, if you crush a piece of paper it will remain crushed unless you straighten it. However, if you squash a piece of foam it will return to its original shape when you stop squashing it. Other words that students may use to describe 'squash' are 'squeeze', 'dent' or 'bend in'. Hard, rigid materials cannot be squashed by hand although they may be squashed by immense pressure.

Flexible materials can be *bent* or folded. Metals can be bent and they will (mostly) retain the bent shape. Paper can be folded (a form of bending) and unfolded. Rigid materials such as glass or stone cannot be bent. Whether a solid material is flexible or not is related to the arrangement of particles in the solid, but students do not need to understand these principles in any depth at this level.

Introduction

- Remind students about the material *clay*, which they were introduced to in Unit 3.5. Recap what the action of *squashing* means: to push material together. Use the examples of sitting on a cushion and squeezing a tube of toothpaste to illustrate this for students. Point to different objects and ask students to say whether they can or cannot be squashed.
- Use a drinking straw and a piece of paper to show that some materials *bend* easily. Demonstrate that a bend can be a curve (as you would get if you tried to bend a plastic ruler) or a fold. Recap the term *flexible*. Explain that this word describes materials that can bend. Point to some items. Let the students take turns to say whether the items are flexible or not.

Teaching and learning activities

- Use the pictures on Student's Book page 48 to start a discussion about how we can change the shape of clay in pottery. Ask questions such as: *What shape is the clay to start with? What can we do to change the shape? What actions would you need to do to make a flat clay disk?* Talk about how raw clay is soft and wet, so its shape

Chemistry • Topic 3 Materials

can be changed, but once the clay is fired (or dried) it becomes hard and you can no longer change its shape by squashing it. Let the students answer the questions orally and then take feedback as a class.

- Show the class Video C5 to demonstrate how metals can be squashed by strong forces.
- Demonstrate squashing an elastic and a non-elastic material. Squash a lump of modelling clay and show the students that it remains squashed when you stop pushing on it. Use a soft ball or a bathroom sponge and gently squash it then let go; it should return to its original shape again. Ask: *What do we call this property when a material is stretches or squashed and goes back to its first shape?* Elicit the term *elastic* and ask the students to give some more examples of elastic and non-elastic materials. This will provide a useful recap of Unit 3.8. Address any misconceptions.
- Explain to the students that you are going to squash some different objects to see how easy it is to squash them. Give each group a selection of objects to test. First, ask them to predict whether they will be able to squash each object. Then allow them time to test each one and record their results in the table on Workbook page 41. Discuss findings as a class. Ensure that all students fully understand the concept of squashing materials.
- Turn to Student's Book pages 48–49. Work through the questions as a class. Include all students and allow them time to think and to ask questions. Ensure all students understand the concept of bending materials.

Graded activities

1 Give each pair of students some coloured pens and a large sheet of paper. Ask them to make a poster to show things they can find at school or at home that will change shape if they squash or bend them. They can include some of the objects they have encountered in this lesson but should also add some of their own ideas. Encourage them to add labels with the name of each object and the materials.

2 Ask the students to work in pairs. Give each pair a sheet of paper. They should make a list of five materials and rank them in order depending on how easy they are to bend. Point out that the list should be of *materials* rather than *objects*. They do not have to actually do the test but they should be able to discuss ideas for how it could be done and correctly rank the materials. Take feedback and discuss as a class.

 Discuss the scenario as a class. Ask students to turn to Workbook page 42 and to draw or write their predictions and why they think this. Explain that they are now going to squash the two different kinds of balls to see if their predictions were correct. Give the students time to test the different balls and record their results. Invite students to feedback their findings to the class. Complete the activity by asking: *What do you think would happen if you tried the test on a table tennis ball or a golf ball?* If students are unfamiliar with these types of balls, you could show them an example or picture and describe their properties. Share ideas as a class and discuss any gaps in understanding.

Consolidate and review

- Ask the students to make a display of materials that can and cannot be squashed. Encourage them to find pictures in old magazines or photos for their display.
- Ask: *How many times can you bend (fold) a piece of paper in half then in half again?* Let the students guess the answer and then try it. Most pieces of paper cannot be folded more than seven times (and young students will struggle to do that many). This is a fun activity that often surprises them. (The world record is 12 folds in case anyone asks.)
- Show the class Slideshow C3. Let them identify all the bent materials on the bridges.

Differentiation

■ All of the students should be able to name and draw some objects that can be squashed or bent. Some students will develop more comprehensive lists.

● Most of the students should be able to list five materials. Some of the students will be able to rank them correctly and suggest ways this could be tested.

▲ Some of the students should be able to predict that the tennis ball will squash and the cricket ball will not. However, this does depend on them knowing that the tennis ball is soft and the cricket ball is hard. Circulate, offering support as necessary.

Chemistry • Topic 3 Materials

3.10 Stretching and twisting

Student's Book pages 50–51

Chemistry learning objective

- *Changes to materials:* 1Cc.01 Describe how materials can be changed by physical action, e.g. stretching, compressing, bending and twisting.

Thinking and working scientifically

- *Scientific enquiry: purpose and planning:* 1TWSp.02 Make predictions about what they think will happen.
- *Carrying out scientific enquiry:* 1TWSc.04 Follow instructions safely when doing practical work; 1TWSc.05 Collect and record observations and/or measurements by annotating images and completing simple tables.
- *Scientific enquiry: analysis, evaluation and conclusions:* 1TWSa.01 Describe what happened during an enquiry and if it matched their predictions.

Resources

- Workbook pages 43, 44 and 45
- Slideshow C4: Stretchy materials

Classroom equipment

- modelling clay and elastic bands
- balloons
- pieces of cloth
- large sheets of paper
- rulers and pencils

Key words

- stretch • twist • elastic

Scientific background

A *stretch* is a pull. A *twist* is a pull with a turn. Certain solids can be stretched and twisted by an applied force. Some of these objects will remain stretched when you stop applying the force. For example, if you stretch a piece of clay it will not return to its original shape when you remove the stretching force. However, if you stretch an elastic band, assuming it has not broken, it will return to its original shape afterwards.

The students should be familiar with the term *elastic* from earlier units in this topic and should recall that an elastic material is one that returns to its original shape after a squashing, stretching or twisting force is removed. Non-elastic materials either do not stretch when pulled or do not return to their starting shape when the force stops.

Introduction

- Introduce the key words for this unit: *stretch* and *twist*. Write the words on the board and point out how each one is spelled and pronounced.
- Use an elastic band and a lump of modelling clay to demonstrate a 'stretch' and a 'twist'. Point out that the elastic band goes back to its original shape when you stop stretching and/or twisting it; the clay does not. Ask: *What do we call this property of the elastic band?* Elicit the term *elastic* and add this key word on the board. Recap that this means it returns to its original shape. The clay is not elastic. (Scientifically, we say it is plastic, but the students do not need to know that at this stage.)

Teaching and learning activities

- Turn to Student's Book pages 50–51. Discuss the questions with the class. If you have a balloon, use it to show what happens as you blow it up and let the air back out.
- Take students' suggestions for names of other things that can be stretched and discuss which of these go back to their original shape once the stretching force has been removed (the elastic items). Ensure that all students fully understand the concept of stretching materials and the property of elasticity.
- Ask students to show how they would twist (wring) a piece of cloth. Give a piece of cloth to each student and ask them to twist it. Ensure that all students fully understand the concept of twisting materials: it involves a stretch and a turn at the same time.

Chemistry • Topic 3 Materials

- Show the class Slideshow C4 and ask the students identify the stretchy materials.
- Turn to Workbook page 43. Explain the scenario to the class: five materials were tested by adding a heavy weight to the end of each one. Each piece of material was the same length at the start of the test. The results show how much each material stretched. Make sure the students understand this and then let them complete the questions in pairs. When they have finished, discuss answers as a class.

Graded activities

1 Give each student a lump of modelling clay and allow them time to experiment with stretching it and twisting it into different shapes. This should be a free and open activity to allow students to experiment with stretching and twisting and observing the results of these actions. Ask them to record their results by drawing the shapes they made on paper. Students can then work in pairs to compare their shapes.

2 Let the students work independently to complete this activity. Using the modelling clay, they should try to stretch it into the longest pencil shape that they can without breaking it. Tell them to draw their clay shape, measure its length and record their results on Workbook page 44. They should then see how many times they can twist their shape before it breaks and record their results. Feedback as a class. Ask: *Who made the longest shape? Who could twist their shape the most times?*

3 Ask the students to work in pairs to complete the activity on Workbook page 45. Make sure the students record their predictions before they do the tests. They should then test to find out whether the thin pencil shape or the thick pencil shape can be twisted more times before it breaks. Invite pairs to share their findings with the class. Did they match their predictions? Ask and answer questions to establish the reasons for the results.

Consolidate and review

- Ask the students to make a list of five materials that will change shape if they are subjected to one of the actions that have learned in this lesson and in Unit 3.9. Let them exchange lists and have their partners say what action could be applied to change the shape of the objects on the list.

Differentiation

■ All of the students should be able to shape the clay by stretching and twisting it. Most students will realise that the shapes you can make are related to the shape of the clay that you started with.

● Most of the students should be able to stretch and twist the shape until it breaks. Some may struggle with measurement, as it requires them to measure both pieces of the broken shape and you may need to help them do this. A few students may realise that the thickness or thinness of the original pencil shape affects the results.

▲ Some of the students will make an accurate prediction, but most will simply guess. However, once they have done a test, more students should be able to predict based on their results. A few students will realise that a thin shape can be twisted gently more times than a fat shape, especially if it is also longer. Encourage them to try and explain why this is the case.

Chemistry • Topic 3 Materials

Science in context

3.11 Uses of science

Student's Book pages 52–53

Chemistry learning objectives

- *Materials and their structure:* 1Cm.01 Identify, name, describe, sort and group common materials, including wood, plastic, metal, glass, rock, paper and fabric; 1Cm.02 Understand the difference between an object and a material.
- *Properties of materials:* 1Cp.01 Understand that all materials have a variety of properties; 1Cp.02 Describe common materials in terms of their properties.

Science in context skills

- 1SIC.02 Talk about how science explains how objects they use, or know about, work.

Resources

- Workbook pages 46 and 47
- Slideshow C5: Materials
- PCM C7: Which is the most waterproof?
- PCM C8: My favourite bridge

Classroom equipment

- selection of objects made from different materials
- magnifying hand lenses
- cut out pictures of things made from wood and stone, glue
- yoghurt pots with a hole in the base, teaspoons or syringes for measuring out small fixed amounts of water, paper towels, sand timers, samples of a selection of materials, such as cotton, wool, sugar paper, newspaper, aluminium foil, plastic (samples large enough to cover the base of the yoghurt pot and go part way up the side)

Key words

- material • object • properties

 Supervise students when they are handling breakable materials such as glass. They should not handle any objects with sharp edges or points. Supervise the students when they use glue.

Scientific background

Materials are the matter from which things are made. Materials may come from animals (for example, fur, wool, silk), from plants (for example, cotton, straw, linen, wood) or from minerals (for example, rock, stone, metal). Some materials are made from other materials (for example, paper, concrete, iron). Many materials are mixtures of different substances.

At this age children may still find it hard to differentiate between the object and the material from which it is made. *Objects* are things that are made from materials. In this topic, students have learned the names of some common materials and their different properties. Now they will use their acquired knowledge and skills to talk about how science explains how familiar objects work.

In Stage 2, students will learn why materials are chosen for specific purposes based on their properties, and that materials can be tested to determine their properties. This unit gives students an introduction to thinking about how materials science explains how objects work. They do not need to understand the detail at this stage.

Introduction

- Recap the key words: *material*, *object* and *properties*. Ensure that all students are clear on the difference between a material and an object that is made from it. Briefly review what students learned about their senses in Topic 2. Elicit the names of the five senses and sense organs to check that students recall these correctly. Tell the class that they are now going to use their senses to explore some different materials.
- Ask students to look at the pictures on Student's Book page 52. Answer the first question as a class. Then get the students to work in pairs to talk about what the things look and feel like. If necessary, prompt them with questions such as: *What does the chair look like? Is it hard or soft?* Take feedback as a class.

Teaching and learning activities

- Show the class Slideshow C5, of different materials. Review that materials are what objects are made from and that there any many different types of materials in the world. Take one of the materials from the slideshow, for example the

Chemistry • Topic 3 Materials

piece of wood. Ask questions about its features: *What colour is it? What does it look like? What does it feel like? What does it sound like if you tap it?* Focus on things the students can easily identify using their senses.

- Show the students a range of objects made from different materials. Allow them time to explore them. You could provide magnifying hand lenses to allow students to examine them more closely. Then ask some questions: *Do you know what the object is? Can you describe what it feels like? What is it made from?* Students can reply using: *This is a… It feels… It is made from…* Encourage them to use as many descriptive words as they can, either in their own language or in English, and compile a class list on the board.

- Using the same selection of materials, ask students to think about what other objects are commonly made from each material. Recap that there are different types of some materials, for example, wood, fabric and metal, and that each type has different properties. Lead the discussion onto how we use each material and why each object is made from that material, using students' knowledge of properties. Ask: *Why do you think it is made from plastic? Would it be good to make it from metal? Why not?* Discuss ideas as a class.

- Make sure the students understand the questions on Student's Book page 53. Ask them to discuss their answers in groups, then invite students to share their ideas with the class.

Graded activities

1 Give the students some pictures of various things made from wood and stone. They should choose one thing made from each material and stick it on Workbook page 46. They should then write the name of the object and describe the features of the material it is made from. Circulate, offering support to any students who may need help in completing this activity.

2 Ask students to complete the activity on Workbook page 47, looking at different materials used in the classroom. They should draw and label pictures of three things, say how each one is used and describe the properties of the material. Encourage students to select different materials, for example, a wooden table, a metal shelf, a brick wall. Help them to identify the materials that have been used to make the objects and to choose the correct adjectives to describe them. For example, *a chair* can be made from *wood* that is *strong* and *smooth*. Point out that they can use the class list on the board for ideas.

3 Tell the class they are going to test some materials to find out how waterproof they are. Demonstrate the test on PCM C7. Put a piece of material over the bottom of a yoghurt pot in which a hole has been made. Stand the yoghurt pot on a paper towel. Add a fixed amount of water (from a teaspoon or syringe) and after a certain time look to see if any water has soaked onto the paper towel. Ask students to carry out the investigation in groups. They need to predict which fabric will be the most waterproof and to explain why. Discuss students' reasons. Ask: *What properties does an umbrella need to have?* Take feedback, ensuring that *waterproof* is one of the properties listed. Ask: *Which material would make the best umbrella?*

Consolidate and review

- Ask students to research bridges around the world and to choose their favourite. Using PCM C8, they draw a picture of their chosen bridge, describe the materials it is made from and the properties of those materials.

- Let students work in groups to make a fun list of materials that are *not* suited for making certain objects, such as a paper coat or a metal hat.

Differentiation

■ All of the students should be able to use everyday adjectives to describe the features of the materials.

● Most of the students should be able to accurately draw and describe their chosen things. For those who need some prompting, circulate, asking questions to guide their thinking.

▲ Some of the students should be able to make predictions about the outcome of the investigation and then follow instructions to find out if their predictions were correct. Offer support and guidance to students who may not be at this level yet.

Chemistry • Topic 3 Materials Consolidation

Consolidation

Student's Book page 54

Chemistry learning objectives

- *Materials and their structure:* 1Cm.01 Identify, name, describe, sort and group common materials, including wood, plastic, metal, glass, rock, paper and fabric; 1Cm.02 Understand the difference between an object and a material.
- *Properties of materials:* 1Cp.01 Understand that all materials have a variety of properties; 1Cp.02 Describe common materials in terms of their properties.
- *Changes to materials:* 1Cc.01 Describe how materials can be changed by physical action, e.g. stretching, compressing, bending and twisting.

Resources

- Workbook page 48
- Video C6: Sportspeople
- Topic quiz sheets C1 and C2

Classroom equipment

- paper, scissors, glue
- selection of pre-cut pieces of fabric
- example(s) of sports equipment

Looking back Topic 3

- Use the summary points on Student's Book page 54 to review the key things that the students have learned in the topic. Ask questions such as: *How many different materials can you name? In what ways are the properties of metal similar to wood? In what ways are they different? Can you name some different objects made from plastic and paper? What two ways can we use to sort objects into groups? How can we change the shape of a piece of fabric?*
- Show the class Video C6, which shows people taking part in different sports. Discuss the different objects and different materials shown. If possible, have some examples of specialist sports equipment available for the students to examine. Then ask the students to complete the activity on Workbook page 48. They should name the material(s) that each object is made from and then think of all the properties that each one has. Ask questions such as: *Which object needs to be flexible? Is the object light/heavy/hard/soft?* Finish by reviewing that materials can have more than one property. This activity will show you how well the students have understood the topic.

How well do you remember?

You may use the revision and consolidation activities on Student's Book page 54 as a paired class activity. If you are using the activities to assess individual learning, have the students work on their own to complete the tasks in writing. If you are using them as a class activity, you may prefer to let the students do the tasks orally. Circulate as they discuss the questions and observe the students carefully, to see who is confident and who is unsure of the concepts.

Some suggested answers

1. Students' own collages.
2. Wood: trees, table, bench, stool, paper napkins, newspaper, basket; fabric: clothes, tablecloth, basket cover, hair ribbons; glass: glasses, jug; plastics: glasses, jug, plates, ball; metal: forks; stone: rocks; combined materials: digital camera and basket; the balloon was stretched.
3. Cricket bat: hard and strong, wood; tennis ball: light and flexible, fabric or plastic; weights: heavy and strong, metal; trainers: light and flexible, plastic, fabric or leather.

Consolidation

Consolidation and reinforcement of the students' understanding of the topic can be undertaken using Topic quiz sheets C1 and C2. This can be completed in class or as a homework task.

Topic quiz sheet answers

Sheet C1

1. Students' own answers

Sheet C2

1. Students' own answers
2. Students' own answers
3. shirt, sponge
4. transparent, smooth, easy to clean
5. True

Chemistry • Topic 3 Materials Student's Book answers

Student's Book answers

Pages 32–33 (3.1)
1. Students' own descriptions
2. Students' own observations as to which features, e.g. shape, colour and size, are the same and which are different.
3. Students' own suggestions; the simplest ways would be based on shape, size and colour.
4. Students' own suggestions; size and colour can be used to sort them but also pattern, transparency, etc.
5. Metal: screw, paperclips, can; wood: plane, paper and card, book; plastic, helicopter, bottles, calculator

Pages 34–35 (3.2)
1. Glass, clay (stone), wood and hair/plastic (brush bristles). Students' own observations; for example: the glass is transparent; the brush is soft, the pot is hard but not as hard as the glass; the pot is orange, the brush is brown, etc.
2. The stone for the pavement.
3. To provide a firm and hard-wearing surface.
4. No. The bottom has split because the bag is not strong enough to carry the shopping.

Pages 36–37 (3.3)
1. Students' own answers
2. Answers might include: shop windows, car windscreens, sunglasses, car headlights.
3. Students' own answers
4. Students' own answers
5. Waterproof does not let water in. Absorbent does let water in.

Pages 38–39 (3.4)
1. For wood, answers might include: wardrobe, chair, bed, bedside cupboard. For plastic, answers might include: toys, clock.
2. Answers might include: duvet cover, pillow, cushion, clothes, teddy bears, carpet, etc.
3. For metal, answers might include: car, bus, umbrella frame, bicycle, motorbike, lamp post, etc. For glass, answers might include: windows, car windscreen
4. Answers might include: umbrella, raincoat, car, bus, house, etc. You could pour water on to it to see if it goes through or is stopped by the material.
5. They are strong, hard wearing and waterproof.

Pages 40–41 (3.5)
1. Students' own answers
2. Plastic, glass, metal; strong and hard wearing.
3. Plastic, fabric
4. Answers might include: lightweight, cheap, keeps the food warm/cool.
5. Students' own answers

Pages 42–43 (3.6)
1. Answers might include: fabric, wood, glass, metal, water, paper, etc.
2. Goldfish bowl, mirror, window
3. Desk, books, etc.
4. Students' own answers
5. Students' own answers

Pages 44–45 (3.7)
1. Wool, silk, cotton
2. Students' own answers
3. Answers might include: lightweight, transparent, strong, soft, easy to mould, etc.
4. It is lightweight, flexible, cheap, easy to mould, etc.
5. Students' own answers

Pages 46–47 (3.8)
1. Flag – fabric; comb – plastic; spoon – metal; bottle – glass; bag – paper; ball – plastic; ruler – wood; jumper – fabric; elastic band – rubber; cup – cardboard and plastic
2. Fabric, paper, cardboard, rubber
3. No
4. Yes
5. Flatten it out (stretch it), squash it further, twist it

Pages 48–49 (3.9)
1. Working the clay to shape it and make pots.
2. The actions of the man – squashing and stretching it.
3. No, the finished pots are hard and can no longer be squashed or stretched.
4. The nail
5. Metal is strong so it cannot be bent by hand; you need a hammer to bend it (with a bigger force).

Pages 50–51 (3.10)
1. It stretches and blows up into a much larger shape.
2. The air escapes and the balloon gets smaller/returns to its original shape.
3. Students' own answers
4. Students' own answers
5. Students' own actions

Pages 52–53 (3.11)
1. Chair – wood; keys – metal; bucket – plastic; glass – glass; hat, scarf and gloves – wool
2. Students' own answers
3. Students' own answers
4. For example, bridge, cars, lamp posts
5. Metal is strong, hard and waterproof.

Physics • Topic 4 Forces and sound

Thinking and working scientifically

4.1 Thinking and working scientifically

Student's Book pages 56–57
Thinking and working scientifically
- *Carrying out scientific enquiry:* 1TWSc.02 Use given equipment appropriately; 1TWSc.03 Take measurements in non-standard units; 1TWSc.04 Follow instructions safely when doing practical work; 1TWSc.05 Collect and record observations and/or measurements by annotating images and completing simple tables.

Resources
- Workbook pages 49 and 50
- PCM P1: How far does it go?

Classroom equipment
- several footballs
- chalk or other markers
- several toy cars
- metre rules or pieces of string
- large sheets of paper, coloured pens or pencils

Key words
- investigation • instructions • equipment
- measurements • observations • record • push

 Remind the students to be careful when they are rolling balls and toy cars to each other.

Skills and connections

Practical work and investigations will be a fundamental part of science lessons throughout the students' school life. Exploring through investigation allows students to answer science questions and find out if predictions were correct. Understanding how to follow instructions safely and use equipment appropriately should be established as good practice from the very beginning. This will provide students will a good foundation for approaching all practical tasks they encounter in the future.

As part of their investigative work, students will be expected to take measurements in a range of ways, both using measuring equipment such as rulers and scales, and by measuring in non-standard units such as foot lengths, as they did in Unit 2.7.

Another key aspect of any practical work is being able to accurately collect and record observations and measurements. Students will need to be able to do this in a range of ways, including annotating diagrams and completing simple tables.

The skills of following instructions and recording results are likely to be applied in other subjects across the curriculum. They may also encounter instructions at home, for example for building a toy. The context for this thinking and working scientifically unit is movement and pushes, which students will learn more about later in the topic.

Introduction

- Remind students of one or two investigations they have done in earlier units, for example in Unit 3.3 (properties of materials). Write the key words *investigation*, *instructions* and *equipment* on the board and point out how each word is pronounced. Relate these words to the investigations you have just discussed.
- Invite students to describe situations where they follow instructions at school or at home, for example sports activities or cooking recipes.
- Explain to students that instructions provide step-by-step notes for what to do in practical work and how to do it. Ensure that students understand the difference between *instructions* and *equipment*. Remind the class that investigations can help us to answer science questions and to find out whether our predictions were correct or not.
- Ask the students to look at the picture on Student's Book page 56. As a class, discuss what the children are doing in the picture (cutting, sticking, making a collage). Then ask the students to discuss in pairs what instructions they think the children have and what these might say to help keep the children safe. Ask questions to guide them: *What do the children need to do first? What should they do next? What do they need to be careful about? How could they get*

Physics • Topic 4 Forces and sound

hurt? Take feedback and discuss as a class. Write a final class list of instructions on the board.

Teaching and learning activities

- Push a toy car to make it travel across the classroom floor. Ask the students to talk to their partner about why it moves. Students will learn about pushes as a force in Unit 4.4, so for now any ideas about movement and your hand are acceptable. Some students may begin to use the word *push*. Introduce this key term to the class and add it to the board.

- Provide each group of students with a football. They should roll the ball to each other at about the same speed. Ask: *What do you feel?* Then ask them to push the ball harder and softer. Ask: *Is it different? How is it different?* Explain that a bigger push makes the ball go further.

- Explain to students that they will often need to take *measurements* and/or make *observations* when they do investigations. Write these key words on the board and point out how each word is pronounced. Ensure that students understand the difference between them: one is a quantitative value (whether in standard or non-standard units) and one is a qualitative description. Explain that it is important to *record* results and add this key word to the board, too.

- Make sure the students understand the questions on Student's Book pages 56–57. Discuss them as a class.

Graded activities

Students complete all of the activities in mixed ability groups. Differentiation should be through the level of support the students receive as they work on each activity, as well as by outcome (please see guidance in the 'Differentiation' box right).

1 Give each student a copy of PCM P1. Ask them to work in pairs to roll a football, gently at first and then more strongly. Guide the students to think of a simple way to record the distance each ball rolls, for example by using a chalk mark or other marker. Once they have a way of recording the distance, they should roll the ball first with a small force and then with a big force. Ask them to record their results by drawing and annotating pictures on PCM P1. Ask: *Which ball went further? Why do you think it went further?*

2 Give each group of three students an identical toy car each. Tell them that you would like one of them to give their car a small push, one to give a medium push and one to give a big push. Ask them to work in groups to predict which car will move the furthest. They should then plan an investigation to find out. Ask the students to write their instructions on Workbook page 49.

3 Set up a straight racetrack for the toy cars. Mark a start line and a finishing line, about a metre apart. Explain to the students that their cars must start behind the start line and that they can only give their car one push. Ask the groups to carry out their investigation to find out which car moves the furthest. Remind one student to push with a small force, one with a medium force and one with a big force. Students should see which car moves the furthest and measure the distances. Show them how to use metre rules or pieces of string. Ask the students to record and explain their results on Workbook page 50, and to compare them with their predictions. Discuss their findings as a class.

Consolidate and review

- Give each group a large sheet of paper and some coloured pens. Ask them to design a safety poster with instructions for crossing the road.

Differentiation

■ All of the students should be able to describe the differences between the two pushes and draw reasonable representations of their results, with just a little help.

● Most of the students should be able to work together, asking and answering questions to clarify their thinking and to predict the outcome of the investigation. Most should be able to explain their reasoning.

▲ Some of the students should be able to work collaboratively and support each other in order to follow instructions and carry out the investigation. Some of the students should be able to accurately record their results in tabular form.

Physics • Topic 4 Forces and sound

4.2 Movement

Student's Book pages 58–59

Physics learning objective
- *Forces and energy:* 1Pf.01 Explore, talk about and describe the movement of familiar objects.

Thinking and working scientifically
- *Carrying out scientific enquiry:* 1TWSc.01 Sort and group objects, materials and living things based on observations of the similarities and differences between them; 1TWSc.05 Collect and record observations and/or measurements by annotating images and completing simple tables.

Resources
- Workbook pages 51 and 52
- PCM P2: Animal movement
- Video P1: Construction machines and vehicles
- Video P2: Animals moving

Classroom equipment
- large sheets of paper, pictures of animals, glue

Key words
- move • movement

 Supervise the students to make sure they do not put anything in their mouths. Supervise the students when they use glue.

Scientific background

Students learned about growth and senses in earlier topics. In this unit, they will look at the different types of movement of familiar things, both living and non-living.

Animals' movements depend on the shapes and sizes of their bodies (legs, wings, fins) and where they live (land, water or both). Animals' movements can also depend on how they get their food and how they protect themselves.

The human body is adapted to stand erect, to walk on two feet and to use the arms to carry and lift. Humans are capable of a wide range of movements, from walking to swimming to jumping.

Non-living things cannot move by themselves. A force needs to be applied to cause the movement and the force can also determine the type of movement, for example whether a ball rolls or spins. Students will learn about pushing and pulling forces later in this topic.

Introduction

- Recap what the students learned in Topic 1 by asking: *What can humans and animals do? Why do animals need to move?* (to find food and water, to find shelter and to stay safe from predators)
- Write *move* on the board and introduce the second key word: *movement*. Jump up and down on the spot a few times. Ask: *What am I doing?* Explain that jumping is a type of movement.

Teaching and learning activities

- Ask the class to look at the pictures on Student's Book page 58. Ask them to discuss their answers to the questions in groups, then take feedback as a class, encouraging them to try to use as many different movement words as possible. Write a class list on the board. Elicit words such as *run, spin, swing, jump, swim,* etc.
- Show the class Video P2, of different animals moving. Allow the students to talk about what they have seen. Ask: *Which animals move slowly? Which move quickly? Which animals swim? Which animals crawl?* Introduce more movement words as necessary, such as *slither,* and add these to the list on the board. Play the video again. Invite the students to mimic how the animal moves.
- Lay out a set of cards cut out from PCM P2. Ask a student to pick up a card, show it to you and then role-play the animal. The rest of the class must try to identify the animal. When all the cards have been used, ask the class to sort them into groups based on the different ways that the animals move. Establish that all animals can move but they do so in different ways.
- Ask the students to stand, spaced well apart. Say: *Show me how many different ways you can move.* Let them hop, jump and wave their arms,

Physics • Topic 4 Forces and sound

etc., to show you. Ask: *Which parts of your body were you using? Do people move in the same way as animals? Do young animals move in the same way as their parents?*
- Show the class Video P1, of different construction machines and vehicles. Write *lifting, tipping, digging, pushing, mixing, turning, moving forwards* and *moving backwards* on the board. Ask the students to describe the movements of the machine or vehicle using these words.
- Make sure the students understand the questions on Student's Book page 59. Ask them to discuss their answers in groups, then take feedback.

Graded activities

1 Show the class Video P2 again, pausing after each animal. Ask: *Which animals move slowly? Which move quickly? Which animals run? Which animals jump? Which animals can move in more than one way? Which ways can they move?* Add any new movement words to the list on the board.

2 Give each pair of students a large sheet of paper, some pictures of animals and some glue. Ask them to sort the pictures into groups based on the ways that the different animals move. They should then divide their paper into different sections, one for each movement type (swim, crawl, walk, etc.) and stick the pictures to make a poster.

3 Talk to the students about the animals that live near your school, e.g. humans, birds, insects. Remind them to think about those that live in trees or in water, as well as those on land. Then take the class on a walk around the school grounds. Tell them to do a survey of the different ways they see animals moving. They should record their findings on Workbook page 51. Ask: *Do animals have only one way of moving? Can a bird move in more than one way?* Discuss how to classify animals such as birds (walk or fly or both?) and agree a consistent way to record their movement. Depending on ability, the students can display their results as a bar chart. Explain that this is a useful method for recording data.

Consolidate and review

- Use Workbook page 52 to consolidate the teaching and to check the students can identify things that move in different ways.

Differentiation

■ All of the students should be able to recognise some of the familiar ways in which the animals move and name the different types of movement.

● Most of the students should be able to sort a selection of animal pictures into groups based on type of movement. Circulate, reminding students that some animals will be able to fit into more than one group as they can move in more than one way.

▲ Some of the students should be able to follow instructions and accurately record data using first-hand observations. Recording their findings in tabular form and then converting them into a bar chart are both useful exercises for practising a method for recording data.

BIG CAT 🐾

Students who have read *Collins Big Cat: I Can Do It!* will notice the different ways that animals move, and will see how children copy these movements.

Physics • Topic 4 Forces and sound

4.3 Pushing and pulling

Student's Book pages 60–61

Physics learning objective
- *Forces and energy:* 1Pf.02 Describe pushes and pulls as forces.

Thinking and working scientifically
- *Scientific enquiry: purpose and planning:* 1TWSp.01 Ask questions about the world around us and talk about how to find answers.
- *Carrying out scientific enquiry:* 1TWSc.01 Sort and group objects, materials and living things based on observations of the similarities and differences between them.

Resources
- Workbook pages 53 and 54
- PCM P3: Pushing and pulling

Classroom equipment
- cardboard box
- some heavy objects, such as books
- selection of toys that work by pushing or pulling
- two large hoops for sorting toys into
- wooden or plastic building brick (or similar)
- toy car with string attached for pulling it
- coloured pens or pencils

Key words
- force • push • pull

Scientific background

For an object to move, there needs to be a force acting on it. A *force* can most simply be described as a *push* or a *pull*. Squashing, stretching, bending and twisting are also types of forces that students learned about in Topic 3. Forces cannot be seen but they can be felt. The students will need the opportunity to try pushing and pulling for themselves to fully understand the concept. The effects of forces can be seen in the resulting motion: the students will see what happens to toys when they are pushed or pulled.

Pushing and pulling are basic ways of making things move. Even when we are using complicated devices, at the simplest level the devices work by pushing or pulling actions. Gears in machines or cars work by pushing, and trucks and trailers work by pulling. Railway systems work by pulling, as all the trucks or carriages are joined to each other and are pulled by the railway locomotive. Force should not be confused with energy or power, both concepts that will introduced in later stages.

Introduction

- Ask two students to come to the front of the class. Ask one student to push gently on the other student. Ask the student who was pushed to push back gently on the first student. Ask the class: *What has just happened?* Take answers and discuss students' ideas.
- Ask another pair of students to the front and tell one of the pair to pull the other gently. Ask the students: *Can you describe how this is different from what the other two students were doing?*
- Ask the students to stand in pairs. Say that, when instructed, one of them should gently push the other. Then the same student in each pair should gently pull the other. Repeat the process, but change the active member of the pair. The purpose of this activity is to set the scene for the unit.
- Introduce and discuss the key words: *force, push, pull*. Write them on the board and point out how each word is spelled and pronounced. Students will probably be familiar with the words push and pull from everyday use, but force is likely to be new to them. Tell the class that they are now going to find out more about pushes and pulls and what a force is.

Teaching and learning activities

- Ask the class to give examples of pushes and pulls in the world around them. Talk about how they know these are pushes and pulls. What questions do they need to ask to find out the answer?
- Start with a wooden or plastic building brick. Ask: *Can we make this brick move? What must we do to it?* Encourage the idea of pushing the brick as well as picking it up, throwing it, dropping it, etc. Replace the brick with a toy car that has a piece of string attached to it. Ask: *What can we do to make this move?* Encourage the students to talk about how to pull the car.

Physics • Topic 4 Forces and sound

- Ask the class to look at the pictures on Student's Book page 60. Ask: *How is the boy making the ball and the car move?* Establish that pushes and pulls are types of forces and that a force can make something move.
- Ask the students to look around them and to think of things in the classroom that they regularly move by pushing (e.g. the door, the light switch, etc.) and by pulling (e.g. the door, drawers, etc.). Demonstrate one of each type. Ask the students to work in groups to make a list. Share the students' ideas and make a class list on the board. Encourage students to use the words *push* and *pull* when they describe the actions.
- Discuss how we push and pull things. For example, we push a wheelbarrow and a vacuum cleaner, but we pull suitcases on wheels and some toys. Invite students to try pushing, and then pulling, a heavy box of books across the floor. Before they start, ask: *Which is easier, pushing or pulling? What can you do to find out the answer?*
- Give each group a selection of toys. Encourage the students to talk about how to make each toy work, and then to make each toy work, before sorting the toys into those that need a push and those that need a pull. Provide two hoops to help with the sorting. Ask: *Were there any toys that were difficult to sort?*
- Make sure the students understand the questions on Student's Book pages 60–61. Establish that some things will work with either a push or a pull, and that we usually use our hands and arms to push and pull things.

Graded activities

1 Ask the students to complete the activity on Workbook page 53. They should colour all the pictures that show a push. Circulate, asking questions as necessary to guide them: *What does the boy do to make the toy truck move? What do his hands do on the handle? Does the truck move away from him or towards him? What do we call the force that makes an object move away from you?*

2 Ask the students to work in pairs. Tell them to choose an object that works by being pushed and one that works by being pulled. They should think about what the objects do and how they work. Allow each pair time to plan how they are going to mime the actions to show how their chosen objects work. They should then perform the mimes to the class. Ask the rest of the class to try and guess the objects and whether they understood what actions were being mimed. Discuss each one as a class.

3 Give each student a copy of PCM P3. Ask them to think about toys that need to be pushed or pulled to make them work. Tell them to draw pictures of the toys that they have thought of in the spaces on PCM P3, in the left-hand column for push toys and in the right-hand column for pull toys. Circulate, looking at the pictures that students have drawn and asking them how each toy works. Clarify any misconceptions.

Consolidate and review

- Use Workbook page 54 to consolidate the teaching and to check that the students can describe the force you use to make one push toy and one pull toy work.
- Allow the students to work in groups to investigate how a cardboard box filled with heavy objects can be moved by pushing or pulling. They can then remove some of the objects and compare how easy the box is to push and to pull.

Differentiation

■ All of the students should be able to correctly identify the things that show a push, with little prompting.

● Most of the students should be able to work collaboratively, sharing ideas and thinking creatively about how best to mime the actions to show an object that works by pushing and an object that works by pulling. Remind them to exaggerate their actions so that it is clear what they are showing.

▲ Some of the students should be able to draw a reasonable representation of their chosen toys and then describe how each one works using the words *push*, *pull* and *force*. If some students struggle to do this independently, display the push and pull toys from earlier in the lesson and allow them to choose from this selection to complete the activity.

BIG CAT

Students who read *Collins Big Cat: Pushing and Pulling* will recognise all the different ways there are of pushing and pulling things to make them move.

Physics • Topic 4 Forces and sound

4.4 Pushes and pulls

Student's Book pages 62–63

Physics learning objective

- *Forces and energy:* 1Pf.02 Describe pushes and pulls as forces.

Thinking and working scientifically

- *Scientific enquiry: purpose and planning:* 1TWSp.02 Make predictions about what they think will happen.
- *Carrying out scientific enquiry:* 1TWSc.02 Use given equipment appropriately; 1TWSc.03 Take measurements in non-standard units; 1TWSc.04 Follow instructions safely when doing practical work; 1TWSc.05 Collect and record observations and/or measurements by annotating images and completing simple tables.
- *Scientific enquiry: analysis, evaluation and conclusions:* 1TWSa.01 Describe what happened during an enquiry and if it matched their predictions.

Resources

- Workbook pages 55 and 56
- Video P3: Waterfalls
- PCM P4: Making a waterwheel

Classroom equipment

- non-motorised model car
- pictures of windmills
- paper, pieces of expanded polystyrene, cocktail sticks, paper sails, troughs of water or access to sink
- coloured pens or pencils
- stiff cardboard (e.g. corrugated), plates to draw round, scissors, sticky tape, skewers, source of water (tap or syringe)
- selection of five balls of different masses and sizes (make sure that the students will be able to move them by blowing on them)
- chalk or tape, tape measure, hand-held fan (battery operated)

Key words

- force • water • wind

 Supervise the students when they use scissors. Supervise the students carefully when they push cocktail sticks into polystyrene, and when they push skewers through their cardboard waterwheels. (You may prefer to do these steps for them.)

Scientific background

We use different types of *forces* every day. Pushes and pulls can help us to do useful jobs.

Water can push with a large force. The faster the water flows and the bigger the volume of water, the greater the pushing force and the more movement it can produce. The waterwheel is an example of humans using the pushing force of water to do a job.

Wind can also provide a pushing force on an object. The stronger the wind, the greater the pushing force and the more movement it can produce.

At this stage the students have not been taught about gases. They may think that air is 'nothing'. It is important that the students feel the force of moving air for themselves so that they can understand this lesson. In the context of sailing boats, a bigger sail means that the force of the wind acts over a larger surface area, and this provides a greater overall force.

Introduction

- Take a toy car and put it on a table. Ask a student to come to the front and make the car go. Ask the class: *What did they do to make the car go? Did they push it or pull it?* Review the previous lesson and establish that pushes and pulls are examples of forces.
- Ask the class to look at the pictures on Student's Book page 62. Ask: *Where is pushing being shown? Where is pulling being shown?* Identify which pictures show pushes and which show pulls. Ensure that the students know the difference between the two actions. Explain that we can use these forces to do useful jobs and that they are now going to look at some examples.

Teaching and learning activities

- Show the class Video P3, of some waterfalls. Ask the students to describe what the water is doing.

Physics • Topic 4 Forces and sound

Discuss that the falling water produces a very large force. Tell the students that we can sometimes use the force of falling water to do useful jobs. Ask the students to look at the picture of a waterwheel on Student's Book page 63. Ask: *What makes the wheel turn?* Explain that waterwheels were once built to use the force of water to do useful jobs such as grinding flour.

- Ask the students to copy your actions. Hold a hand up in front of your face, with the palm facing towards you. Blow gently on your hand. As the students blow on their hands, ask: *What do you feel?* Then hold a piece of paper in front of your face and ask the students to copy you again. Blow on the paper. As they blow on their paper, ask: *What happens when you blow on the paper?* Show the students some pictures of windmills. Ask: *What makes the sails turn? Where does the force come from to make the sails turn?*

- Look at the pictures of sailing boats on Student's Book page 63. Discuss what is making the boats move. If wind is given as the answer, ask individual students if they can improve on this by using some science words. An improved answer would be that the force of the wind is pushing on the sails. Make sure the students understand the question. Explain, in simple terms, that the boat with the bigger sail will move faster because more wind can push on it, creating a bigger force.

- Let the students make their own boats. Use a small piece of polystyrene for the boat, a cocktail stick for the mast and a piece of paper for the sail. Using a water trough or sink, the students test their boats by blowing gently and then harder. The students should describe what happens, using the words *blow*, *hard*, *gently*, *move*, *fast*, *slow*.

Graded activities

1 Ask the students to complete the activity on Workbook page 55. Ask: *Have you all circled the same things? Were there any things that you could not identify? Which things needed pushing or pulling to make them work?* Ask the students to explain their answers.

2 Give each group a copy of PCM P4 and a range of materials (stiff cardboard, plates to draw round, scissors, sticky tape, skewers). Read out and demonstrate the instructions, then offer help as necessary as the students construct their own waterwheels in groups. Test the waterwheels in front of the whole class. Encourage the students to make comments and suggestions. Ask them to consider their own waterwheels, and those of others, and suggest ways of improving the waterwheels to make them work better. Ask: *What can we do to make the waterwheel turn faster? Why does this happen?* Discuss, in simple terms, that if the water flows faster or we use more water, the pushing force will be bigger and the wheel will move faster.

3 Show students five different balls. Ask: *Which ball do you think will be the easiest to move by blowing? Which do you think will be the hardest to move by blowing?* Let the students make predictions on Workbook page 56. Then ask them to decide how they could test their predictions. Challenge the students to test the balls and rank them in order of easiest to most difficult to move by blowing. For some groups provide chalk or tape to mark the distance travelled by each ball. Allow others to use a tape measure. Demonstrate the test using a hand-held battery-operated fan to provide a constant stream of air. Students should record their results on Workbook page 56.

Consolidate and review

- Ask the students to work in pairs to describe how a waterwheel and a windmill work. This will show how well they have understood the unit.

Differentiation

■ All of the students should be able to correctly identify the things that work by pushing or by pulling.

● Most of the students should be able to work together to follow instructions to make their waterwheel with some support.

▲ Some of the students will be able to carry out the investigation independently, recording their results and asking and answering questions to clarify their thinking.

Physics • Topic 4 Forces and sound

4.5 Floating and sinking

Student's Book pages 64–65
Physics learning objective
- *Forces and energy:* 1Pf.03 Explore that some objects float and some sink.

Thinking and working scientifically
- *Scientific enquiry: purpose and planning*: 1TWSp.02 Make predictions about what they think will happen.
- *Carrying out scientific enquiry:* 1TWSc.02 Use given equipment appropriately; 1TWSc.04 Follow instructions safely when doing practical work; 1TWSc.05 Collect and record observations and/or measurements by annotating images and completing simple tables.
- *Scientific enquiry: analysis, evaluation and conclusions:* 1TWSa.01 Describe what happened during an enquiry and if it matched their predictions.

Resources
- Workbook pages 57 and 58
- Digital resource P1: Will it float?
- PCM P5: Float or sink?
- PCM P6: Will it float?
- PCM P7: Making paper boats

Classroom equipment
- aluminium foil
- scissors, and glue or sticky tape
- basins half-full of water
- variety of objects that will float or sink, e.g. keys, glass marbles, plastic toys, polystyrene pieces, pencils, paperclips, toy car, wooden bricks, plastic cup, toy duck

small weights of equal mass, e.g. small coins
Key words
- float • sink

 Remind students to take care with water spillages. Mop up any spillages immediately to avoid slipping. Supervise the students when they use scissors and glue.

Scientific background

Floating is usually taken to mean a solid thing resting on the surface of a liquid. This can be a boat supported by water or a duck swimming.

An object's density is a measure of its mass relative to its volume. Objects *float* in water if their density is less than that of water, and they *sink* if they are denser than water. For example, a stone has a greater mass than an equivalent volume of water and so it sinks, whereas a piece of polystyrene foam has much less mass than an equivalent volume of water, and so it floats.

An object's shape can affect whether or not it floats. A steel ship has a very large mass, but its shape means that it contains a large amount of air, reducing its overall density to a point where a volume of water the same size as the ship weighs more than the ship itself. Another example is a ball of clay, which will sink, but a model canoe made from the same amount of clay can float because it pushes more fluid out of the way in relation to its weight. Hollow objects are able to float better than solid objects.

Introduction

- Let the students complete Digital resource P1 to say if the different objects will float or sink in water. Ask the students: *Can you tell me the name of something that floats? What about something that sinks?*
- Ask the class: *Have you ever been on a boat?* Discuss the students' experiences, paying attention to the size of the boat and what it was made of. Ask: *Do you know why a boat floats?* Explore students' ideas about floating and sinking. Tell them that they are now going to learn more about this.

Teaching and learning activities

- Ask students to look at the pictures on Student's Book page 64. Ask: *What do the pictures show? What is floating? What is sinking?* Answer the questions as a class and discuss answers. At this

Physics • Topic 4 Forces and sound

stage, the students do not need to understand the science behind it but they should know that when an object is supported by a liquid (usually on the surface of water) it is *floating*. Write the two key words *float* and *sink* on the board and point out their pronunciation.

- Do a class demonstration with a bowl of water and a collection of objects to test floating. Take a metal object, such as a spoon, and show it to the students. Ask: *Will it sink?* Take a wooden or plastic brick and ask the same question. Test the objects. Ask the students: *Were the results what you expected?* Take feedback and check that they understand the words *float* and *sink*.

- Give each pair of students a copy of PCM P5 and ask them to cut out the objects and stick them in the right place on PCM P6: either floating on the surface of the water or lying at the bottom of the bowl. Check answers as a class and clarify any misconceptions.

- Ask students to look at the picture of the ships on Student's Book page 65. Ask: *Why don't they sink?* Take answers and discuss ideas about what makes an object float or sink. Introduce the fact that the type of material and the shape of the object are important to whether an object can float or not.

- Many students will think that heavy objects sink and light objects float. (You can refer back to Unit 3.2 where they learned about the properties *heavy* and *light*.) Make a shallow, flat-bottomed boat shape out of aluminium foil. Show students that it floats, even though it is made of metal. Squash the foil tightly into a ball and show the class that it now sinks. Ask: *What about the shape made it do this?* Establish that as well as material and weight, shape can also affect whether something floats or sinks: for example, the large metal ship is heavy but its shape (making it full of air) means it can float.

Graded activities

Students complete all of the activities in mixed ability groups. Differentiation should be through the level of support the students receive as they work on each activity, as well as by outcome (please see guidance in the 'Differentiation' box right).

1 Give each group a selection of objects. Ask them to predict whether each one will sink or float. They should then test the objects and record their findings on Workbook page 57. Discuss why the objects behaved as they did. Ask: *How are the objects that sank similar to each other? How are the objects that floated similar?*

2 Ask students to work in groups to build the four boats in the picture, using PCM P7 as a guide, and then test them to see which ones float the best. Ask: *Which boat floats the best? Why do you think it floats best?* They should be able to see that the boats that contain the most air (due to their shape) float the best.

3 Ask students to pick the most successful boat and see how much weight it can take before it sinks. They can do this by adding coins or marbles, one at a time. Initially they should find that the boat holds the weights but sits lower in the water. Tell them to add more until the boat sinks. Ask them to record their investigation by drawing and labelling their boat floating and the boat when it has sunk, on Workbook page 58. Discuss the features of the boat that made it good at holding the weights.

Consolidate and review

- Let the students play with some toys and water to experiment with things that float and sink.
- Ask: *If you were trapped on an island and had to make a boat to get home, what would you make it from? What shape would it be?* This activity will give a good indication of students' understanding of this unit.

Differentiation

■ All of the students should be able to predict and then test the different objects to see which will float and which will sink.

● Most of the students should be able to construct the boats by following the instructions supplied and say which shape floated best.

▲ Some of the students should be able to explain why the chosen boat was best (shape). All of the students should be able to select the most successful boat and test it by adding weights.

Physics • Topic 4 Forces and sound

4.6 Listen carefully

Student's Book pages 66–67

Physics learning objective
- *Light and sound:* 1Ps.01 Identify different sources of sound.

Thinking and working scientifically
- *Carrying out scientific enquiry:* 1TWSc.05 Collect and record observations and/or measurements by annotating images and completing simple tables.

Resources
- Workbook pages 59 and 60
- PCM P8: Sound source survey

Classroom equipment
- small bell
- picture of a musical band
- plain paper or notebooks
- coloured pens or pencils
- six identical glass bottles filled with different amounts of water

Key words
- sound • source

 If you take the students on a walk around the school grounds, ensure they are safe and that they stay together. Be aware of any students who may have a hearing impairment, and adapt the lesson accordingly. Remind students to take care with water spillages. Mop up any spillages immediately to avoid slipping.

Scientific background

A sound is made when something moves backwards and forwards very quickly. We usually call this *vibrating*. The scientific term for it is oscillating. The frequency of sound is measured in Hertz (Hz). A musician would call a note with a frequency of 40 Hz a low note; a note with a frequency of 90 Hz would be called a high note.

If you blow across the top of a bottle you can hear a note. The less liquid there is in the bottle, the lower the note will be; it is the air inside the bottle that is oscillating. If you want to make a louder sound you have to put in more effort, e.g. by hitting a drum skin harder. By putting in more effort into plucking a string or hitting a drum, you make the size of the oscillations bigger. This is called the amplitude. A loud sound is made by an oscillation with a large amplitude. Different instruments have different sound qualities; this is known as the tone of the instrument.

At this stage, the students will learn about and identify different sources of sound; they do not need to understand how a sound is made. (This will be covered in more detail in Stage 5 of this course.)

Introduction

- Show the students a picture of a band as a talking point. Ask the students to describe what they can see. Remind the students about their five senses, which they learned about in Topic 2, and ask them to name each sense and the sense organ associated with it. Introduce the key word *sound* and elicit that sounds are heard when they enter the ear. Ask the students to imagine what sounds they would hear if they were listening to the band in the picture. Tell them that they are going to learn all about sound in this topic.

- Ask the students to look at the top picture on Student's Book page 66, and discuss question 1 as a class. Next, let the students look at the bottom picture and discuss questions 2 and 3 in small groups. After a few minutes, take feedback from the groups and see if they all identified the same things that do not make sounds.

- Introduce the key word *source*. Invite the students to suggest what this might mean. Ascertain that the source of a sound is where it comes from. This will help you assess their prior knowledge of the subject.

Teaching and learning activities

- Ring a small bell. Ask the class: *Can you tell me how you know that the bell rang?* Take some responses. Ask: *What part of your body are you using to listen? Where are your ears? What other sounds do you hear with your ears?* Take some responses.

Physics • Topic 4 Forces and sound

- Form two equal groups. Number the students in each group, so that everyone has a partner. Match each number to an animal, for example, the number ones are cats, the number twos are goats, etc. The students must keep their number and animal secret. When you say *Start!* they must make the sound of their animal and try to find their partner, based on hearing the animal sounds. The game ends when all the students have found their partners. Take feedback, asking: *What did you have to do to find your partner? Was it easy? Why?*
- Arrange six identical glass bottles in a line, with different amounts of water in each. Show the students how to blow across the top of a bottle to make a sound. Let the students blow across the tops and describe the sounds. They should then try tapping the bottles with a pencil. Ask: *Do all the bottles produce the same sound? Can you describe the sounds? Are the sounds the same when you tap the bottles and when you blow over them?* Let the students discuss the answers. Accept any reasonable descriptions. You could write a list of the descriptive words on the board.

Graded activities

1 Ask the students to turn to Workbook page 59 and to look at the activity. Make sure the students understand the words *bang, buzz, click, hiss* and *splash* before they begin. Ask the students to complete the activity on their own. Circulate and help any students that are having difficulty matching the sounds to the sources of the sound.

2 Ask the students to sit quietly in the classroom. Ask: *How many different sounds can you hear? What is the source of each sound?* Encourage them to make a list (by writing or drawing) and to classify the sounds as being either natural or human-made. Ask: *Why do you think this?* Let the students share their answers and write a list on the board of all the sounds they heard. Identify and discuss any differences the students may have observed when identifying a sound as being either natural or human-made.

3 Divide the students into groups of four or five. Take them for a walk around the school grounds to listen for different sounds. They can use copies of PCM P8 to record the sounds that they hear and their sources, or they can write their own lists in their notebooks. Encourage the students to use a variety of descriptive sound words, either in English or in their own language, for example *rumbling, whistling, rushing, pattering, crashing*. Once back inside the classroom, ask the students to draw a map of the school grounds (or to copy a map of the grounds you have sketched on the board). They should mark on the map the place where they heard each sound and add a label to say what the source of the sound was, for example: *whistling/tweeting – bird; rumbling – a car/truck*. They should also say whether the sound was natural or human-made.

Consolidate and review

- Use Workbook page 60 to consolidate the teaching. Make sure the students understand the words *beep, bang* and *click* before they begin.
- Let the students work in pairs, at opposite ends of a long table. Ask one student to put their head on the table, with their ear pressed against it. Ask the other student to tap the table very gently. The first student should count how many times the second student taps the table. Let them swap roles.

Differentiation

■ Some of the students will need help reading the words and matching them to the pictures. Students working at a higher level may like to suggest some alternative descriptive words that could be used.

● Most of the students should be able to write the different sounds they can hear.

▲ Some of the students will confidently create their own lists of sounds and sources of sounds without needing to use PCM P8. They will be able to accurately copy a map of the school grounds from the board and add locations and labels with little or no help.

Physics • Topic 4 Forces and sound

4.7 What made that sound?

Student's Book pages 68–69

Physics learning objective

- *Light and sound:* 1Ps.01 Identify different sources of sound.

Thinking and working scientifically

- *Scientific enquiry: purpose and planning:* 1TWSp.02 Make predictions about what they think will happen.
- *Carrying out scientific enquiry:* 1TWSc.01 Sort and group objects, materials and living things based on observations of the similarities and differences between them; 1TWSc.05 Collect and record observations and/or measurements by annotating images and completing simple tables.
- *Scientific enquiry: analysis, evaluation and conclusions:* 1TWSa.01 Describe what happened during an enquiry and if it matched their predictions.

Resources

- Workbook page 61
- PCM P9: Musical instruments
- PCM P10: Sound word cards
- PCM P11: Blow, pluck or strike?
- Audio clip P1: Sounds familiar
- Audio clip P2: What is it?
- Video P4: Music

Classroom equipment

- a selection of different musical instruments for students to explore, if possible (or pictures of musical instruments): include string, wind and percussion

Key words

- sound • source

 Be aware of any students who may have a hearing impairment, and adapt the lesson accordingly.

Scientific background

Sound is made by vibrations that travel through the air to our ears. Stringed instruments make a sound when their strings are plucked or strummed because the strings vibrate and the vibrations are transferred to the air. The sound of a wind instrument comes when we blow air into it, causing the air inside to vibrate. Percussion instruments vibrate when they are struck. The human voice is made by breathing air over our vocal cords, causing them to vibrate; it is like a cross between a stringed and a wind instrument.

Musical instruments work by amplifying vibrations. The body of the instrument resonates and amplifies the noise, making it louder.

At this stage, the students will learn about and identify different sources of sound; they do not need to understand how a sound is made. (This will be covered in more detail in Stage 5 of this course.)

Introduction

- Remind the students about the sounds they heard in previous lesson. Ask them to sort the sounds into two groups, guiding them to say *natural sounds* and *human-made sounds*. Ask the students to look at the pictures on Student's Book pages 68–69. Elicit that natural sounds can be made by the wind or rain. Discuss how humans and animals can make sounds. Ask: *How many different ways can you think of that humans make sounds?* (speaking, singing, moving, etc.)

- Ask the students: *Do you know anyone who plays a musical instrument? What kind of instrument is it? Can you describe how it makes sounds?* If appropriate, show the students Video P4, then ask them to describe and explain how each instrument is played.

- Return to Student's Book pages 68–69 and answer questions 1–3 as a class. Elicit that musical instruments can be plucked, blown into or struck in order to produce a sound. Let the students answer questions 4 and 5 in their groups.

Teaching and learning activities

- Give the students copies of PCM P9 and ask them to draw a line to match each musical instrument with the action that is required to produce a sound.

Physics • Topic 4 Forces and sound

- Hand out a set of word cards cut from PCM P10 to each group. Read each word with the class, asking them to repeat it after you and making sure the students understand what each word means. Say the name of a musical instrument out loud, either in English or in the students' own language (you could also show pictures of the different instruments). The students should hold up the correct word card to match the way that instrument is played.

Graded activities

1. Ask the students to look at the picture on Workbook page 61. They should circle all of the sound sources. Let them swap books with a partner to see if they have circled the same sources. Still in pairs, ask the students to act out a sound from the picture and let their partner point to the source of the sound. Come together and ask the students to say some words that describe the sounds in the picture, for example *loud*, *quiet*, *noisy*. Write these on the board. Tell the students they will be learning more about loud and quiet sounds later in this topic.

2. Tell the students that they must sit very quietly and listen to some sounds. Play Audio clip P1 of some familiar sounds, all the way through. Repeat the audio clip, but this time ask the students to name the source of each sound. Go through the answers with the class, identifying each source of sound. (1. Applause, 2. Bell, 3. Rooster crowing, 4. Thunder, 5. Water boiling, 6. Walking on gravel) Ask the students to say where they might hear each sound. Ask: *Is this a natural sound or a human-made sound?*

3. Give groups of students three different musical instruments to explore (each group should have a wind, string and percussion instrument). Ask them to predict what sound each one will make. They should then use the instrument to make a sound. Ask: *Was the sound the same as you predicted? If not, why not?* As a whole class, ask the students to sort the musical instruments into three groups. Lead the students to sort them into wind, string and percussion. You can use copies of PCM P11 to help guide students. Ask: *Why have you chosen these groups?* (blow, pluck, strike/hit) *Can any of the instruments be put into more than one group? Can the groups be sorted into smaller groups?* (for example, drums, bells, shakers) Ask students to demonstrate how each instrument works by blowing, plucking or striking it.

Consolidate and review

- Play Audio clip P2 of some unfamiliar sounds. (1. Bubbles underwater, 2. Sawing wood, 3. Camera taking photos with film, 4. Helicopter, 5. A sheet of glass smashing, 6. Human heart beat). Ask the students to identify the sound sources. Help them as necessary by acting out some clues.

- Play a game of 'I can hear with my little ear, something that sounds like...'. The students should name the source of the sound.

- Ask the class the following questions and discuss their responses: *In today's lesson, what did you hear? What did you use to hear these things? Was it difficult to name the sounds from the audio clips? Can you describe the sounds in detail? Use as many describing words as you can.*

Differentiation

■ All of the students should be able to circle the sources of sounds. They will be able to use some descriptive words to describe the sounds, including *loud* and *quiet*, and understand what they mean.

● Most of the students should be able to identify and correctly name the sources of the sounds from the audio clip, with little prompting. Most students should be able to say if the sounds are natural or human-made; students not working at this level may need some additional guidance.

▲ Some of the students will correctly predict the sound each musical instrument will make. They will be able to confidently sort them into three groups and suggest ways of sorting them into smaller groups. Students working at a higher level may begin to make the connection between sound and the vibration of a musical instrument. Most of the students will be able to correctly sort the instruments into three groups using PCM P11 as a guide. A few students may need some prompting in order to predict the sounds each instrument will make and to sort them into three groups.

Physics • Topic 4 Forces and sound

4.8 Loud and quiet sounds

Student's Book pages 70–71
Physics learning objectives
- *Light and sound:* 1Ps.02 Explore that as sound travels from a source it becomes quieter.

Thinking and working scientifically
- *Scientific enquiry: purpose and planning:* 1TWSp.02 Make predictions about what they think will happen.
- *Carrying out scientific enquiry:* 1TWSc.02 Use given equipment appropriately; 1TWSc.04 Follow instructions safely when doing practical work; 1TWSc.05 Collect and record observations and/or measurements by annotating images and completing simple tables.
- *Scientific enquiry: analysis, evaluation and conclusions:* 1TWSa.01 Describe what happened during an enquiry and if it matched their predictions.

Resources
- Workbook pages 62, 63, 64 and 65
- PCM P12: Making a shaker

Classroom equipment
- a selection of different musical instruments
- stopwatch
- materials to make shakers: plastic bottles with lids, funnels, lentils, pasta, small stones, rice, etc., to put into the bottles

Key words
- loud • quiet

 Be aware of any students who may have a hearing impairment, and adapt the lesson accordingly. If the students use the internet, ensure they do so safely and under adult supervision. Supervise the students when they use scissors.

Scientific background

Sound is the movement of energy through a gas, liquid or solid in longitudinal waves. Sound is produced when energy causes an object to vibrate. The more energy is put in, the louder the sound will be, for example hitting a drum, blowing a whistle.

Our voices act like a musical instrument. Air from our lungs is passed over the vocal cords, which vibrate to make a sound. We need to have enough air in our lungs to sing loudly, or at all. Having more breath allows us to sing for longer or more loudly. Singing loudly pushes more air over the vocal chords, but takes more energy.

At this stage, the students will learn that we hear sounds with our sense of hearing and our ears are the sense organ for this. Some students may start to ask questions about how sound travels to the ear; this should be explained in simple terms as the science will be covered in much greater detail in Stage 5 of this course.

Introduction

- Remind the students of the musical instrument investigation they did in the previous lesson. Ask them to describe the sounds that the shakers made. Lead them to use the key words *loud* and *quiet*. Suggest some other words such as *noisy*, and write these on the board.
- Let the students explore a selection of musical instruments. Ask them if they can make quiet sounds and loud sounds using the same instrument. Ask: *What did you do to make the loud sounds?* Answers will vary depending on the instruments, but you should elicit that they have to be struck, blown or plucked harder (with more force). Repeat, asking: *What did you do to make the quiet sounds?*
- Ask: *Is there another way to make the sounds louder or quieter?* Suggest that you could move closer to or further away from the source of the sound. Let the students tell the class about their experiences of hearing loud noises but from far away, for example an airplane, a music concert. Some students may struggle with this concept and say that the sound is quiet. Accept this as an answer at this stage. (Students will learn more about sound in Stage 5 of this course.) Ask the students to look at Student's Book pages 70–71 and answer the questions as a class.

Physics • Topic 4 Forces and sound

Teaching and learning activities

- Tell the students that they are going to investigate how long they can sing a note for. Use a stopwatch to time how long the students can sing for. Now ask them to sing the same note as loudly as they can, and time how long the note lasts. Ask: *Do you think you sang for longer when you were singing louder? Why did you stop singing? Did you run out of breath sooner when you sang louder?* Now time how long the students can sing quietly. Ask them: *Why could you sing for longer when you sang quietly?* Tell the students to breathe out as far as they can, and then try to sing a note as loudly as they can. Ask: *Can you explain why you could not sing?* Tell the students to take a large breath in, and then try to sing a note as loudly as they can. Ask: *What has happened to the volume of your voice?* You could try asking the students to sing a song at different volumes, to see whether singing loudly makes them more tired than singing quietly.

Graded activities

1 Ask the students to look at the pictures on Workbook page 62. Explain that they are going to draw lines from each source of sound to a number. The numbers 1 to 5 represent how loud the sound is, with 5 being the loudest. If necessary, help students by completing the first one as an example. When they have finished, discuss the results. Some children may have ranked the sounds differently to others. Ask individual students to explain to the class why they have put the pictures in a particular order. Encourage them to use words such as *loud, quiet, noisy, low, high, deep*. It might be useful to write some descriptive words on the board for the students to use.

2 Give each student a copy of PCM P12 and the equipment they will need to make a shaker (provide a variety of fillings to put inside the shakers so they are not all the same). Go through the instructions on PCM P12 and then let the students make their shakers. Circulate and offer guidance to any students that find the task difficult. Ask the students to draw a picture of their finished shaker and to describe it, on Workbook page 63. Encourage them to use words such as *loud, quiet, noisy*, etc., to describe the sound that their shaker makes.

3 The students should plan and carry out an investigation using their shakers. Ask them to turn to Workbook page 64. Read through the text with the students and then let them complete the sentences and make their predictions. They can use the descriptive words you have written on the board to help them. Let them test their shakers. They should record their results on Workbook page 65. Ask the students to compare their predictions with their results.

Consolidate and review

- Let the students use their shakers. Ask them to make a loud sound and then a quiet sound. Let them start with a quiet sound and get louder and louder.

Differentiation

■ All of the students should be able to draw lines from the sound sources to the numbers. Most will understand the ranking system being used, although students not working at this level will need additional help with this concept. All of the students should be able to use a variety of descriptive words.

● Most of the students should be able to follow the instructions in the correct order and make a shaker. All of the students will be able to draw a picture of their shaker and describe the sound it makes.

▲ Some of the students should be able to independently write plans and predictions for their investigation in their Workbook. They should be able to explain their reasoning.

BIG CAT

Students who have read *Collins Big Cat: Pam Naps* will be familiar with the idea that sound travels, and will be intrigued by how much noise can go on around Pam before she wakes up.

Students who have read *Collins Big Cat: How to Make a Maraca* may be inspired to make their own musical instruments, to supplement what they have done in the lesson.

Physics • Topic 4 Forces and sound

4.9 Sound and distance

Student's Book pages 72–73
Physics learning objective
- *Light and sound:* 1Ps.02 Explore that as sound travels from a source it becomes quieter.

Thinking and working scientifically
- *Scientific enquiry: purpose and planning:* 1TWSp.02 Make predictions about what they think will happen.
- *Carrying out scientific enquiry:* 1TWSc.03 Take measurements in non-standard units; 1TWSc.04 Follow instructions safely when doing practical work; 1TWSc.05 Collect and record observations and/or measurements by annotating images and completing simple tables.
- *Scientific enquiry: analysis, evaluation and conclusions:* 1TWSa.01 Describe what happened during an enquiry and if it matched their predictions.

Resources
- Workbook pages 66, 67, 68, 69 and 70
- PCM P13: Sound and distance

Classroom equipment
- toy train
- ticking clock
- MP3 player or other portable device to play music
- metre rules or long tape measures
- paper and drawing equipment

Key words
• distance • source • loud • quiet

 Be aware of any students who may have a hearing impairment, and adapt the lesson accordingly. If you take the students out of the classroom, ensure they are safe when they are in the school grounds.

Scientific background

Sounds are produced by vibrations. Vibrations travel away from a source through the air. To hear a sound, the vibrations have to travel from the source of the sound to our ears.

Sound waves spread out from the source, rather like the ripples spreading out on the surface of a pond. Their effect becomes less the further they spread out. By damping the vibrations, the volume of sounds can be reduced. Soundproofing materials work by damping vibrations.

At this stage, the students will learn that as sound travels away from its source it becomes fainter and more difficult to hear; they do not need to understand about vibrations and sound waves. Some students may start to ask questions about how sound travels to the ear; this should be explained in simple terms, as the science will be covered in much greater detail in Stage 5 of this course.

Introduction

- Ask the students if they have ever been to a train station or an airport. Ask: *What did the train coming into the station sound like?* (It was quiet when it was far away and became louder as it approached.) Let the students take turns to model the sound of a train approaching and leaving a station using a toy train. Make sure that the students understand that the train is the source of the sound.

Teaching and learning activities

- Arrange the children in three lines, standing facing forward. Show them the ticking clock and ensure that they can all hear it. Bring forward any children who cannot, until they can all hear it. Ask them to sit and to cover their ears. Ask whether they can still hear the clock. Tell them to uncover their ears and explain that you are going to move the clock and that they should sit down when they can no longer hear it. Move away from the children with the clock, until all the children have sat down. Ask them to explain how the sound changes (gets quieter) as the clock moves further away.

- Ask the students to look at Student's Book pages 72–73. Answer the questions as a class, helping students as necessary.

Physics • Topic 4 Forces and sound

Graded activities

1 Tell the students that they are going to investigate what happens to a sound as you move further away from the source. Take the students into the hall or, if it is a fine still day, outside if possible. Line the students up across the hall/playground and explain that you are going to play some music. The students will have their backs to you and should walk away until they cannot hear the music any more. When all of the students have stopped, measure the distance that they have walked (this does not have to be an accurate measurement; for example, it could be counted in paces by the teacher). Repeat the experiment with some louder/quieter music and compare the distances travelled. Establish that the further away you are from the source of a sound, the quieter it becomes. Ask: *From what distance could you hear a quiet sound?* Ask the students to think about the way sound is used as a signal. Ask: *Has anyone heard a sound that is used as a signal for something that is about to happen? How many examples can we think of?* Examples could include emergency service vehicles, flood warning sirens, the fire alarm in school, etc. Back in the classroom you could ask the students to draw a picture to demonstrate how sound is used to warn us of danger.

2 Ask the students to turn to Workbook pages 66–67. Say that they are going to plan their own investigation to see what happens to a sound as you move further away from the source. Read through the text and make sure the students understand what they are going to do. Let them complete their plan and make their predictions. Go through the plans with the students and make sure they are practicable.

3 Let the students carry out their investigations. They should record their results on Workbook page 68. Back in the classroom they should compare their predictions with their results, and explain their findings on Workbook page 69. Ask: *Which group had the best plan? Why? What would you change?*

Consolidate and review

- Ask the students to imagine they are somewhere where there is a very loud noise, for example at the launch of a space rocket. Ask: *What can you do to protect your hearing?* (move further away from the source of the sound; wear ear defenders)

- Give each student a copy of PCM P13 and ask them to complete the activity in order to assess their understanding of sound and distance. This is not as straightforward as it first appears. The students' answers will open up opportunities for further discussion.

- Ask students to complete the crossword puzzle on Workbook page 70 to consolidate their understanding of the key words from this topic.

Differentiation

■ All of the students should be able to take part in the investigation with little or no help. Most will understand that the sound becomes quieter/fainter the further away they walk from the source. They will all be able to take part in the class discussion about sounds being used as warnings, drawing examples from personal experience or their knowledge of the wider world.

● Most of the students should be able to independently plan an investigation and make predictions in their Workbook, although some will require additional help to read the text. A few students will find this task difficult and will benefit from more structured guidance from the teacher.

▲ Some of the students should be able to carry out their investigations without any help. They will independently record their results in their Workbook and compare their prediction with the outcome. Some students will be able to offer suggestions for improving their plan.

BIG CAT

Students who have read *Collins Big Cat: The Pied Piper of Hamelin* may wonder how all of the rats – and then the children – heard the piper playing. In this unit they learn more about how sound travels.

Physics • Topic 4 Forces and sound Consolidation

Consolidation

Student's Book page 74

Physics learning objectives

- *Forces and energy:* 1Pf.01 Explore, talk about and describe the movement of familiar objects; 1Pf.02 Describe pushes and pulls as forces; 1Pf.03 Explore that some objects float and some sink.
- *Light and sound:* 1Ps.01 Identify different sources of sound; 1Ps.02 Explore that as sound travels from a source it becomes quieter.

Resources
- Workbook pages 71 and 72
- Topic quiz sheets P1, P2, P3 and P4

Classroom equipment
- audio clips of different animals

Looking back Topic 4

- Use the summary points on Student's Book page 74 to review the key things that the students have learned in the topic.
- Ask the students to complete the activity on Workbook page 71. They should make a list of some different ball games and say how you make the ball move in each game. Encourage the students to use the word *force* in their answers. This activity will show you how well the students have understood the pushes and pulls as forces.
- Ask the students to complete the activity on Workbook page 72. They should make a list of some different sounds they can hear at home and write the names in the correct rooms on the picture. This activity will show you how well the students have understood the units on sound.

How well do you remember?

You may use the revision and consolidation activities on Student's Book page 74 as a paired class activity. If you are using the activities to assess individual learning, have the students work on their own to complete the tasks in writing. If you are using them as a class activity, you may prefer to let the students do the tasks orally. Circulate as they discuss the questions and observe the students carefully, to see who is confident and who is unsure of the concepts.

Some suggested answers

1 Picture A: two pushes: the wind blowing the kite, and the boys kicking the ball; Picture B: TV, computer, cat, people, fire, etc.
2 Students' own answers, e.g. basketball: bounce, throw; tennis: throw, hit; golf: hit; rugby: throw, kick; badminton: throw, hit.
3 Answers will depend on the audio clips.
4 Students' own answers

Consolidation

Consolidation and reinforcement of the students' understanding of the topic can be undertaken using Topic quiz sheets P1, P2, P3 and P4. These can be completed in class or as a homework task.

Topic quiz sheet answers

Sheet P1
1 Swims – fish; walks – camel; flies – bird; slithers – snake
2 Bird and worm; car and trailer
3 Pushes; faster
4 Push; force

Sheet P2
1 The pull-along giraffe.
2 C; A; No
3 False; false
4 Students' own answers
5 More water; faster water

Sheet P3
1 bird; car
2 Students' own drawings
3 ear
4 Students' own answers
5 Students' own answers

Physics • Topic 4 Forces and sound Student's Book answers

Sheet P4
1 trumpet - blow; guitar - pluck; drum - strike
2 quiet; gently
3 move further away from the source; wear ear defenders
4 fire alarm; police siren
5 True

Student's Book answers

Pages 56–57 (4.1)
1 The children are cutting and sticking to make a collage.
2 Students' own answers, but these should include instructions to cut, stick and make a picture/collage.
3 Students' own answers, but these should focus on being careful with the scissors and the glue.
4 An investigation to find out which car will move the furthest
5 Measurements of distance; students may suggest standard or non-standard units
6 Students' own suggestions, e.g. how far each car travelled, how big a push was needed to make the cars move, etc.

Pages 58–59 (4.2)
1 Answers might include: animals, orangutan hanging from a tree, orcas swimming, deer running/jumping, boy rolling a snowball, horse running, etc.
2 Answers might include: swinging, swimming, running, jumping, rolling, etc.
3 Sitting, running, swimming, catching a ball (playing).
4 Students' own answers might include: jump, skip, run, walk, crawl, climb, shake, swim, etc.
5 Answers might include: flying, walking, climbing, etc.

Pages 60–61 (4.3)
1 He is pushing it.
2 He is pulling it.
3 Push: trolley, tricycle. Pull: giraffe, doll
4 Yes: you can push or pull the trolley, tricycle and giraffe.
5 Your hands/arms.

Pages 62–63 (4.4)
1 You push the shopping trolley. You pull the plug out.
2 Push: camera button, car. Pull: fishing net, cart.
3 The water pushing down on it.
4 The wind pushing the sails.
5 The blue boat. It has a bigger sail so there is a greater force (from the wind) pushing it.

Pages 64–65 (4.5)
1 Key – metal; coin – metal; rock – rock; trays – metal; bottle – plastic
2 Students' own answers
3 Students' own answers
4 Their shape (hollow) allows them to float.

Pages 66–67 (4.6)
1 Answers might include: talking, shouting, footsteps, etc.
2 Answers might include: traffic, people talking, a road drill, baby crying, ambulance siren, etc.
3 Answers might include: windows, clothes, etc.
4 Answers might include: shouting, banging, engine noises, etc.
5 Answers might include: the man is shouting, the digger is moving, the digger's engine is working, etc.

Pages 68–69 (4.7)
1 Baby, whistle, bell, phone, alarm clock, fire engine.
2 It rings when you shake it. / It vibrates and makes a sound.
3 Blow it. / Push air into it.
4 They are making sounds/music using musical instruments (and they are singing).
5 They are moving/shaking/striking/hitting/tapping the instruments.

Pages 70–71 (4.8)
1 Environment B
2 Environment A
3 Environment A: traffic, people talking, people walking, a road drill, headphones, etc. Environment B: birds singing, wind blowing, etc.
4 Loud sounds: lightning, fireworks, digger. Soft sounds: water, mouse, snake.
5 Loud sound: strike it with more force/harder. Soft sound: strike it with less force/gently.

Pages 72–73 (4.9)
1 The child at the front of the line and the children wearing headphones.
2 The boy at the front is further away from the source of the sound. The two children who are wearing headphones cannot hear the drum.
3 Answers might include: birds singing, airplane, wind, footsteps, talking, singing, etc.

Physics • Topic 5 Electricity and magnetism

5.1 What things need electricity?

Student's Book pages 76–77

Physics learning objective
- *Electricity and magnetism:* 1Pe.01 Identify things that require electricity to work.

Thinking and working scientifically
- *Carrying out scientific enquiry:* 1TWSc.01 Sort and group objects, materials and living things based on observations of the similarities and differences between them.

Resources
- Workbook pages 73 and 74
- PCM P14: Things that use electricity
- PCM P15: Mains electricity or batteries?

Classroom equipment
- pictures of devices, some which use mains electricity and some which use batteries
- old magazines, scissors, glue, sheets of paper

Key words
- electricity • mains electricity • battery • device

 Students must not plug or unplug electrical devices that use mains electricity and should be discouraged from doing so. If you take the students on a walk around the school grounds, ensure that they are safe and that they stay together. Supervise the students when they use scissors and glue.

Scientific background

Electricity results from the movement of tiny particles called *electrons,* which carry an *electric charge*. The flow of these charged particles is called an *electric current*. Power stations 'push' electric current, which is connected to devices by plugging them into a *mains* socket. The electric current provides the energy to make the device work.

Batteries (singular: *battery*) are a source of electrical energy. Different types of batteries have different properties and therefore different uses. The amount of energy a battery can provide depends on two main factors: the battery size and the type of cell. Some batteries are rechargeable, while others can no longer be used once they 'run out'.

Many electrical appliances and devices use mains electricity because batteries are impractical. The devices require high levels of energy to work so the use of batteries would be very expensive, the batteries would need to be too big, or they would run out very quickly.

At this stage, the students do not need to understand how electricity works, but they should appreciate that some devices need electricity in order to work. (In Stage 2 students will learn more about electricity, how it can only flow when a *circuit* is complete, and about electrical safety.)

Introduction

- Use the topic opener photograph on Student's Book page 75 as a talking point. Ask the students to describe what they can see. Let them briefly discuss any initial ideas they may have about lights and electricity, battery. Tell them that they are going to learn about things that need electricity to work.

Teaching and learning activities

- Ask the students to look at the pictures on Student's Book page 76. Write the key words *electricity, mains electricity, battery* and *device* on the board and point out how each word is spelled and pronounced. Answer the questions as a class. Discuss which of the things use mains electricity and which use batteries. Ask questions to guide the students as necessary, such as: *Do we need to plug it into the wall to make it work?*

- Give each group of students a selection of pictures of devices, some which use mains electricity and some which use batteries. Allow the students time to discuss the pictures. Ask: *What makes this object work? Which objects use electricity? Which objects use batteries?* Ask the students to sort the objects into two different groups, depending on their power source. Encourage the students to share their ideas, challenge the ideas of other students, and to ask

Physics • Topic 5 Electricity and magnetism

and answer questions in order to find the correct answers.

- Point to different electrical devices in the classroom and ask: *Does this need electricity to work? Does it use batteries or mains electricity?* Ensure that students understand the difference between the two sources of electricity. Explain that batteries are not used for bigger devices because they would run out very quickly. At this stage, only explain mains electricity in simple terms: that it comes from somewhere else in the country and reaches the wall sockets through cables and wiring that is out of sight. Point out that some batteries can be recharged, whereas others can no longer be used once they run out.

- Direct the students to the pictures on Student's Book page 76. Ask students to work in pairs to answer the questions. Take responses from the class. Ask: *Can you name any other things that need electricity to work?* Invite students to describe what the devices that they name do. Encourage all students to share their ideas. Extend the discussion by asking students what it would be like in the classroom without electricity. Ask: *Which thing would you miss the most? Why?* They will explore the concept of a world without electricity more in Unit 5.3.

- Ask students to complete the activity on Workbook page 73. Ask them to work in pairs and look around the classroom to find things that use electricity. If they can find the items that are shown in the Workbook, they should circle 'yes'. Ask them to then draw any other electrical items that they can see. Invite pairs to share their findings with the rest of the class. Ask: *Do any of the items use batteries rather than mains electricity?*

Graded activities

1 Give each pair of students a copy of PCM P14. Ask the students to match each object that needs electricity to what it is used for. Ask questions to guide their thinking as necessary, such as: *What is the name of this object? What does it do? Where might we find one? Can we use it to wash clothes?*

2 Ask students to complete the activity on PCM P15. They need to identify all the things that are powered by mains electricity and colour them in red, and then identify all those that are battery-powered and colour them blue.

Circulate, offering support as necessary. Ensure that students are able to correctly identify the items before they decide whether each one is mains- or battery-powered.

3 💬 Ask students to work in groups and go on a hunt around the school to find four things that use electricity to work, such as lights, computers, air conditioning, fans. Ask them to draw pictures on Workbook page 74 and write the name of each object underneath the picture. They should then describe what each object is used for. Talk through all the items that the students found. Ask: *What do we use this object for? What things do you have at home that use electricity?* Discuss their ideas as a class and address any misconceptions.

Consolidate and review

- Play a class game of 'I spy', choosing an electrical object in the classroom and asking and answering questions in order to establish what it is. The student who correctly guesses gets to choose the next item, and so on.

- Give the students some old magazines and ask them to cut out pictures of objects that use main electricity and batteries. Tell them to sort the objects into two groups and then stick them on a sheet of paper.

Differentiation

■ All of the students should be able to match each object to its use, with just a little help.

● Most of the students should be able to identify each of the objects shown and then correctly say whether each one uses mains electricity or batteries. For those that need some prompting, circulate asking questions to guide their thinking.

▲ Some of the students should be able to work collaboratively to find four things that use electricity, then draw and describe the items to record their findings. If students need some support, you can remind them of some of the electrical items they have discussed during the lesson.

Physics • Topic 5 Electricity and magnetism

5.2 Exploring magnets

Student's Book pages 78–79
Physics learning objective
- *Electricity and magnetism:* 1Pe.02 Explore, talk about and describe what happens when magnets approach and touch different materials.

Thinking and working scientifically
- *Scientific enquiry: purpose and planning:* 1TWSp.02 Make predictions about what they think will happen.
- *Carrying out scientific enquiry:* 1TWSc.04 Follow instructions safely when doing practical work; 1TWSc.05 Collect and record observations and/or measurements by annotating images and completing simple tables.
- *Scientific enquiry: analysis, evaluation and conclusions:* 1TWSa.01 Describe what happened during an enquiry and if it matched their predictions.

Resources
- Workbook pages 75 and 76
- PCM P16: Magnetic or not?
- Slideshow P1: Uses of magnets

Classroom equipment
- magnet and magnetic material
- large containers, fish shapes, small steel paperclips, pencils, string, small magnets
- range of magnetic and non-magnetic materials, to include steel paperclips, aluminium drinks cans, iron nails, coins, plastic objects, wooden objects
- range of different kinds of magnets: horseshoe, bar, old, new, big, small, etc.

Key words
- magnet • magnetic

 Remind students not to drop any of the magnets.

Scientific background

A *magnet* is usually made from a substance that contains iron. A magnet attracts other *magnetic* materials. Materials that are attracted to magnets include *metals* such as iron, some steels, nickel and cobalt. Not all metals are magnetic. Non-magnetic materials include plastic, wood and glass.

Magnets can be used for making toys, for sorting materials (in aluminium recycling centres, for example), for fridge magnets, and for tools. Big electromagnets pick up metal cars and parts in scrapyards. Very strong magnets (neodymium magnets) are used in computer hard drives. Half of all neodymium magnets are found in computer drives.

At this stage, students do not need to know about the poles of a magnet or about the forces of attraction. The purpose of this unit is to introduce students to magnetism and what happens when magnets approach and touch different materials.

Introduction

- Show the students a magnet and a magnetic material. Start with them apart and then move them together. Repeat this several times.

Ask: *Why do they move towards each other?* Explain that one is a magnet and one is a magnetic material. Tell the students that when you use a magnet you can feel a pulling force when it comes close to an object made from a magnetic material. If the material moves towards the magnet, it is magnetic. Write the key words *magnet* and *magnetic* on the board. Point out how each word is spelled and pronounced.

- Write the word *metal* on the board. Remind students about some of the properties of metals that they learned in earlier units (Topic 3). It would also be useful to recap the concept *pull* as a force, which was covered in Units 4.3 and 4.4.
- Invite individual students to come up to the front of the class and use the magnet to try to pick up objects. As each student has a turn, ask the class: *Is this object magnetic?* When they say yes, or no, develop two separate piles of objects on your desk: those that are magnetic and those that are not magnetic. Tell the class that they are going to learn more about magnets in this unit.

Physics • Topic 5 Electricity and magnetism

Teaching and learning activities

- Play a fishing game. Give each group about ten fish shapes, each with a metal paperclip attached to it, and a pencil with a piece of string tied at one end with a magnet tied to the other end. How many fish can each student catch while their team counts to ten? Ask: *How are you able to catch the fish?* Establish that the magnet is 'pulling' the fish towards it.
- Explain that magnets come in all shapes and sizes. Demonstrate the different magnets that you have on your desk. Encourage the students to think of other magnetic objects. Extend the discussion to talk about where and how magnets are used in everyday objects at home and in the classroom. Make a list on the board. Show students Slideshow P1.
- Ask the students to look at the pictures on Student's Book page 78. As a class, read through the text and answer the questions.
- Make sure the students understand the questions on Student's Book pages 78–79. Ask them to discuss in pairs which objects they think are magnetic and which objects they think are not. Then they should think about how they could test this to find out. Take feedback as a class and address any misconceptions. Establish that all students have understood the concepts by asking them each to form a sentence using the words *magnet* and *magnetic*.

Graded activities

1 Give each student a copy of PCM P16. Ask them to work in pairs to tick which objects they think are magnetic. Take feedback and ensure that all students ticked the correct objects.

2 Tell the students that they are going to investigate some different objects. Remind students that magnetic materials will move towards a magnet and non-magnetic materials will not. Give each group a variety of objects and a magnet. Include metal paperclips, an aluminium drink can, a wooden spoon, a metal spoon, paper, iron nails, coins. Ask the students to test each object and then record their results on Workbook page 75. Ask: *How did you know if the object was magnetic? Can you tell if something is magnetic by looking at it?*

3 Provide each group with a magnet. Ask the students to find four metal objects in the classroom and four non-metal objects. When they find an item, tell students that they should first predict if they think it is magnetic. They can then test each one to find out. Ask them to complete the table on Workbook page 76, listing the objects, the materials they think that each one is made from, and whether their predictions were correct. Once the students have completed the activity, ask: *Can you see a pattern in the results?* Students working at a higher level may begin to observe that only certain metal objects are magnetic (iron and steel). An aluminium drinks can, for example, is not magnetic.

Consolidate and review

- Invite volunteers to select various objects in the classroom. The rest of the class should raise their hand to answer 'yes' when the volunteer asks: *Is this magnetic?* Each object can then be tested with a magnet to find out the correct answer.

Differentiation

■ All of the students should be able to recognise that objects that are magnetic.

● Most of the students should be able to recognise that objects that move towards a magnet are magnetic materials. They will be able to record their results with little or no help.

▲ Some of the students should be able to work collaboratively and support each other in order to follow the instructions. Some of the students should be able to accurately record their results in tabular form. If the students are not at this level yet, you may prefer to do this as a teacher-led class investigation.

Physics • Topic 5 Electricity and magnetism

Science in context

5.3 History of science

Student's Book pages 80–81
Physics learning objective
- *Electricity and magnetism:* 1Pe.01 Identify things that require electricity to work.

Science in context skills
- 1SIC.01 Talk about how some of the scientific knowledge and thinking now was different in the past.

Resources
- PCM P17: No electricity
- PCM P18: Then and now
- PCM P19: Before and after electricity

Classroom equipment
- desk lamp and candle
- scissors
- research materials

Key words
- electricity • device

 Do not allow students to get too close to the burning candle. Supervise the students when they are using the scissors. If the students use the internet, ensure they do so safely and under adult supervision.

Scientific background

For centuries, people used fire for heating and cooking, and simple oil lamps and candles for lighting. Alessandro Volta developed the first practical method of generating electricity in 1800, producing a glowing copper wire. Humphrey Davy created the first lamp in 1802, but it was not until nearly the end of the 1800s when Thomas Edison patented the first commercially successful bulb. The first electricity supply for homes was in 1882.

Modern homes heavily reply on things that use electricity as their source of energy. This electrical energy is changed into different forms, such as light, sound, heat and movement. As technologies have improved, electrical appliances have been superseded by better, more effective and efficient products. Devices decrease in size and cost and increase in sophistication with developments and progress in science. For example, a modern mobile phone has vastly more computing power than the computer than guided *Apollo 11* to the moon.

In some parts of the world, mains electricity cannot be taken for granted and may regularly be cut, or supply can be disrupted due to natural disasters, such as hurricane or earthquakes. People in such countries may use different energy sources to light their homes and cook with.

This unit gives students the opportunity to think about the reliance of the modern world on electricity. It will start to give them an insight into how some of the scientific knowledge and thinking that we have now has changed over time.

Introduction

- Briefly recap what students learned about mains electricity in Unit 5.1. Ask them to think about the area in which they live. Ask: *What things outside need electricity? What things at home and at school need electricity to work?* Encourage the class to think of as many electrical devices as possible and make a list on the board.

- Ask the students to look at the pictures on Student's Book page 80. Answer the first question as a class. This activity will serve as good revision of Unit 5.1. Turn to the next question and ask students to consider what it would be like if they woke up one morning to find that the mains electricity supply had been cut. Ask: *What could you still do as normal? What things would not work? What would you not be able to do?* Discuss how the loss of electricity affects the whole community. The students may not realise that in some parts of the world, this is a regular occurrence, so let them know this is the case.

- Explain to students that scientific knowledge and thinking change over time. We did not always know the things that we know now. Electricity is one such example, and this made life very different in the past. For example, light bulbs have only been in use for about 100 years.

Physics • Topic 5 Electricity and magnetism

Teaching and learning activities

- Give each pair of students a copy of PCM P17. Refer back to the idea of waking up to find that there is no electricity. Ask students to suggest some alternative things that could be used and what for. Ask them to draw and describe their ideas on PCM P17, such as candles (for light), fire (for heat and cooking), hand-held fan (for cooling). Ask: *What would you find hardest to do without electricity?*
- Ask students: *What did people use to light their homes before electricity?* Elicit the use of fire, lamps and candles. Tell the class that you are going to compare a candle with an electric lamp. Ideally do this demonstration in a dark room. Light a candle and invite individual students to come up to try to read by candlelight. Then do the same with a modern desk lamp. Ask: *How easy was it to read by candlelight? Which light was better?*
- Write the key words *electricity* and *device* on the board and point out how each word is spelled and pronounced.
- Ask the students to look at the pictures on Student's Book pages 80 and 81, and to discuss the questions in groups. Take feedback and discuss their ideas as a class. Ensure that the students have all grasped the concept that scientific knowledge and thinking change over time, which in turn changes things around us.

Graded activities

1 Give each pair of students a copy of PCM P18. Ask them to cut out the pictures and match the 'then' and 'now' pictures and descriptions, to show how the things have changed over time. Ask: *What are the main differences between the old and modern devices?* Invite students to share their answers with the class and discuss any differences, changing any mismatched pairings as necessary.

2 Ask the students to complete the activity on PCM P19 in pairs. Explain that the page is divided into two equal sections and that they need to choose two things that were different before and after electricity. Ask the students to draw and label pictures of their chosen items, and then to write some sentences to describe the differences. Invite pairs to share their completed work with the class and discuss ideas. How many different 'before' and 'after' things were chosen as a class? Ask: *Do you think your things are better with electricity?*

3 Remind students that electrical devices change over time. Ask them to work in pairs to choose a device that has changed over time and which they would like to find out more about. They can use a device they have seen in the lesson or another device of their choice; you could write some suggestions on the board to guide their thinking. Help them to do some research using reference books or the internet. Students should make a fact sheet to illustrate their findings. They should include pictures to show how the device has changed and write a sentence to describe each change. Display these on the wall in the classroom and discuss.

Consolidate and review

- Ask each group to briefly outline their thoughts on whether electricity is a good source for lighting. Groups should discuss why and why not.
- Ask students to describe how a school classroom would have been different before the use of electricity.

Differentiation

■ All of the students should be able to match the 'then and now' pictures, with just a little help.

● Most of the students should be able to work together to share ideas, asking and answering simple questions and supporting each other, to draw and describe two things 'before' and 'after' electricity.

▲ Some of the students should be able to work collaboratively, thinking critically and asking sensible questions to extend their knowledge and produce an informative fact sheet. Some students may not be working at this level yet; if so, you may prefer to do this as a more structured activity by offering a frame for them to work to.

Physics • Topic 5 Electricity and magnetism Consolidation

Consolidation

Student's Book page 82
Physics learning objectives
- *Electricity and magnetism:* 1Pe.01 Identify things that require electricity to work; 1Pe.02 Explore, talk about and describe what happens when magnets approach and touch different materials.

Resources
- Workbook page 77
- Topic quiz sheet P5

Looking back Topic 5

- Use the summary points on Student's Book page 82 to review the key things that the students have learned in the topic. Ask questions such as: *What can you remember about electricity? Can you name some things in the classroom that use electricity? Can you name something that uses a battery? What does it feel like when you move a magnet close to a magnetic material? Can you name an object that is magnetic?* Encourage the students to use the key words from the topic in their answers.
- Ask the students to complete the activity on Workbook page 77. Tell them to think about things at home, in their living area, that use electricity. They should then draw pictures and add labels to describe what each one does. This activity will show you how well the students have understood the topic.

How well do you remember?

You may use the revision and consolidation activities on Student's Book page 82 as a paired class activity. If you are using the activities to assess individual learning, have the students work on their own to complete the tasks in writing. If you are using them as a class activity, you may prefer to let the students do the tasks orally. Circulate as they discuss the questions and observe the students carefully, to see who is confident and who is unsure of the concepts.

Some suggested answers
1. There are six things in the picture that use electricity: mobile phone, iron, kettle, toaster, microwave oven, vacuum cleaner. The mobile phone has a battery. It uses mains electricity to recharge the battery.
2. Students' own answers
3. The magnetic material moves towards the magnet and you feel a pulling force.

Consolidation

Consolidation and reinforcement of the students' understanding of the topic can be undertaken using Topic quiz sheet P5. This can be completed in class or as a homework task.

Topic quiz sheet answers

Sheet P5
1. Hair dryer; lamp; TV
2. Students' own answers
3. Paperclip; screw
4. Metal; towards; pull

Physics • Topic 5 Electricity and magnetism Student's Book answer

Student's Book answers

Pages 76–77 (5.1)

1. toaster, lamp, fan, smartphone
2. flashlight, remote-controlled car, TV remote control, smartphone
3. Possible answers: air conditioner, fan, lamp, whiteboard, laptop, calculators
4. Calculator and laptop
5. Laptop

Pages 78–79 (5.2)

1. Students' own answers
2. Connecting toys, holding picture on the fridge door, picking up scrap metal
3. The magnet picking up scrap metal. It needs to be big because the metal it needs to pick up is very heavy.
4. Students' own answers
5. With a magnet

Pages 80–81 (5.3)

1. Possible answers: street lamps, lights in buildings, lamp, TV.
2. Students' own answers
3. Students' own answers
4. Students' own answers
5. Possible answers: the modern computer is smaller/slimmer, has fewer parts, has a built-in mouse, has a bigger screen, is portable.

Earth and Space • Topic 6 Earth and Space

Thinking and working scientifically

6.1 Clean water investigation

Student's Book pages 84–85

Thinking and working scientifically

- *Scientific enquiry: purpose and planning:* 1TWSp.02 Make predictions about what they think will happen.
- *Scientific enquiry: analysis, evaluation and conclusions:* 1TWSa.01 Describe what happened during an enquiry and if it matched their predictions.

Resources

- Workbook pages 78 and 79
- PCM ES1: Moving water
- PCM ES2: Filtering water
- PCM ES3: Testing water filters

Classroom equipment

- cup of water, a coin
- for each pair of students: two cups, one full of water, a jug, a sponge, a pipette
- bottles or cups of clean water
- bottles of dirty water (water mixed with mud, leaves, twigs, etc.)
- 2-litre plastic bottle, scissors, cotton fabric, elastic band or string, cotton wool, washed gravel, washed sand, supply of dirty water (water mixed with mud, leaves, twigs, etc.)
- for each group of students: 1-litre plastic drinks bottle cut in half with the bottom part removed, 10 cotton balls, 5 paper napkins, coffee filter, plastic bag, supply of dirty water (add leaves, soil, sand, etc., to the water to make it dirty), beaker, tray

Key words

• investigation • prediction • results

 Do not let the students drink any of the water that they handle. It is important that the students understand that the water you filter in the experiment will be clean, but that it will not be safe to drink. Always supervise students when they are working with water. Mop up any spills immediately.

Skills and connections

Students learned about making predictions in Unit 1.1, answering the questions *What do you think will happen?* and *Why?* and applying their knowledge and understanding of the world to what is happening in a situation. In this unit, students will focus on describing what happens during an investigation and concluding whether or not the results match their predictions. Being able to analyse and evaluate results versus predictions is a key skill that students need to learn, and one which they will continue to use throughout their education.

The context of the unit is water. Although about 70 per cent of the Earth is covered in water, the vast majority of this (97 per cent) is seawater. Seawater has salt in it so is not suitable for drinking. Fresh water comes from sources such as rivers, lakes and wells, as well as taps and bottles. Clean fresh water is safe to drink but dirty water can make humans and animals ill or even lead to death. Water from taps and bottles is usually clean as it will have been processed. Water from wells comes largely from rainwater that has been filtered through the layers of rock to form groundwater. Because it has been filtered, it is relatively clean, but it may still need to be disinfected before drinking.

When filtering water, if the pores in the filter material are too large, the water will not be adequately filtered. If they are too small, they may block up and not allow water to pass through. The students do not need to know this level of detail at this stage, but they will carry out an investigation to introduce them to the idea of filtering and how well substances can filter out different types of 'dirt'.

Introduction

- Remind the students of the earlier work you have done on making predictions, for example in Unit 1.1 (plants). Write the key words *investigation*, *prediction* and *results* on the board and point out how each word is spelled and pronounced.
- Hold a coin over a cup of water and ask the students to predict what will happen when you drop it into the water. Remind them to give reasons and use the sentence structure:

Earth and Space • Topic 6 Earth and Space

I think…because… This will serve as a useful recap of Unit 1.1 and refresh students' memories of what making a prediction involves.

- Drop the coin into the cup. Ask: *What happened? Was your prediction correct?* Remind the class that it does not matter whether a prediction is correct or not; the important thing is that students try to give sensible ideas for what they think the outcome might be. Tell them that in science, we often make a prediction and then do an investigation to find out whether the prediction was correct or not. Say that they are now going to practise doing this.

Teaching and learning activities

- Give each pair of students a copy of PCM ES1, two cups (one filled with water), a small jug, a sponge and a pipette. Ask: *Which way of moving water from one cup to the other will be best? Which will spill the most water?* They should number the methods 1–4 to show their predictions. Allow the students time to test each method and describe what happened. Discuss their findings as a class. Were their predictions correct?
- Ask the students to look at the pictures on Student's Book pages 84–85. Discuss the questions as a class. Elicit that humans need water to stay alive and that drinking dirty water can make us ill.
- Tell the students that most of the Earth's surface is covered in water. (They will learn more about this in Unit 6.2.) Ask students to name places where we can get fresh water. (If no one mentions it, suggest a river.) Give each pair of students a bottle or cup containing some clean water and another containing some dirty water. Let them examine both. Say: *The dirty water is like the muddy water that comes from a river. What can we do to clean the water to make it safe to drink?* Explain that water from most sources must be cleaned before we can drink it. Tell students that they are going to learn about how to make clean water.
- Make sure students understand the questions on Student's Book pages 84–85. Discuss the answers as a class.

Graded activities

Demonstrate the filtration experiment to the class, using the instructions on PCM ES2. Point out the names of the different pieces of equipment as you use them.

1 Ask students to label the picture of the filter equipment on Workbook page 78. Refer them to the experiment they have just watched. Check answers as a class.

2 Before you do the demonstration, ask the students to predict what they think will happen when you pour the dirty water through the filter. Write their predictions on the board. Pour a glass of dirty water through the filter. Ask: *What can you see? Does the water look cleaner?* Allow them to discuss this in groups and then take feedback as a class. Ask: *What happened? Was your prediction correct?* Explain in simple terms that the different materials help to take out different things from the dirty water.

3 Give each group of students the resources they need to construct their own water filters. They should follow the instructions on PCM ES3 to make a model water filter. Before they use the filter, ask them to write their predictions on Workbook page 79. The students should then test each filter material, using their bottles of dirty water, and record their results. Ask: *What happened? What did you find? Were your predictions correct?* Ask students to try to explain how the filters stop large solids from getting through.

Consolidate and review

- Give the students a selection of different sized and shaped containers and some water. Let them pour water from one to the other, trying not to spill any water or have any left over.

Differentiation

■ All of the students should be able to correctly label the water filter equipment.

● Most of the students should be able to follow the demonstration and make a sensible prediction for the outcome.

▲ Some of the students should be able to predict the outcome of the investigation and work collaboratively, with little help.

Earth and Space • Topic 6 Earth and Space

6.2 Our planet Earth

Student's Book pages 86–87

Earth and Space learning objectives

- *Planet Earth:* 1ESp.01 Know that Earth is mostly covered in water.
- *Earth in Space:* 1ESs.01 Know that Earth is the planet on which we live.

Thinking and working scientifically

- *Scientific enquiry: purpose and planning:* 1TWSp.01 Ask questions about the world around us and talk about how to find answers.

Resources

- Workbook page 80
- Video ES1: Earth
- Slideshow ES1: Oceans, rivers and lakes
- PCM ES4: Water on Earth
- PCM ES5: Ocean in a jar

Classroom equipment

- globes
- glass of fresh water and glass of salty water
- clean water-tight jars, sand, water, shells, other 'ocean' items, blue food colouring
- round balloons, paper mâché, paints

Key words

- Earth • planet • water • land

 Always supervise students when they are working with water. Mop up any spillages immediately. Make sure students do not consume any of the water. If the students use the internet, ensure they do so safely and always under adult supervision.

Scientific background

Our Solar System consists of eight *planets*, which orbit the star we call the Sun: Mercury, Venus, *Earth*, Mars, Jupiter, Saturn, Uranus and Neptune. Pluto was previously classed as a planet but in 2006 it was reclassified as a dwarf planet by the International Astronomical Union.

Water is a natural resource and is essential for all known life. Water covers over 70 per cent of the Earth's surface and is present in solid, liquid and gaseous forms. The remaining 30 per cent of the Earth's surface is land, which is composed of rock and soil. Land comes in a wide variety of types, including deserts, grasslands and rainforests, each with their own features.

The terms *ocean* and *sea* are often interchanged. There are five oceans on Earth: the Atlantic, Arctic, Indian, Pacific and Antarctic (or Southern). Seas are smaller than oceans and there are several dozen seas on Earth. Oceans and seas contain salty water. Rivers are freshwater systems that flow into a sea, ocean, lake or other rivers. Lakes are surrounded by land. They contain fresh water.

Introduction

- Use the topic opener on Student's Book page 83 as a talking point. Ask the students to look at the picture of the Earth taken from space and to describe what they can see. Give each group a globe and ask them to compare it to the picture. Let them briefly discuss in groups and then take ideas as a class. This will give you a good indication of students' prior knowledge for this unit.

- Hold up a globe and explain that it is a model of the *planet* we live on. Ask: *Do you know what our planet is called?* Elicit that it is called *Earth*. Point out your country on the globe. Tell the class that *water* is a key feature of the Earth. It covers about three-quarters of the Earth's surface. The rest of the Earth's surface is *land*. Write the key words on the board and point out how each word is spelled and pronounced.

- Ask students look at the picture on Student's Book page 86. Ask: *What is this a picture of? What can you see? Which parts are water and which are land?* Explain that the land is different colours depending on the landscape, for example dry desert areas will be brown whereas lush forested areas will be green.

- Show the students Video ES1. Pause it at key points for the students to discuss what they can see. Show the Atlantic Ocean centred (to show how big it is) with Africa, Europe and the Americas in view. Show the Pacific Ocean

Earth and Space • Topic 6 Earth and Space

centred (to show how huge it is). Show your country centred and point out the surrounding countries and large areas of water.

Teaching and learning activities

- Ask the class: *Where can we find water in nature?* Write a list of the students' suggestions on the board. Discuss the different sources of water, such as oceans, seas, rivers and lakes.
- Show the students two glasses of water, one with fresh water and one with salty water, with all the salt dissolved. Ask: *Are these the same?* Explain that they are different. Tell the students that one is saltwater and the other is fresh water. Stress to the students that, although they look the same, the types of water are different. Ask: *Where can we find large amounts of saltwater?* Tell the students that there are other differences between the types of water that we find on our Earth. (You may like to refer back to Unit 2.5.)
- Show the class Slideshow ES1, of oceans, seas, rivers and lakes. Ask: *How are they similar? How are they different?* Explain the features such as whether they contain fresh or saltwater and whether they are enclosed by land. Then ask the students to draw and label a body of water of their choice on PCM ES4. You can refer them to the list on the board for ideas, if necessary.
- Ask the students to look at the pictures on Student's Book pages 86–87. Ask them to discuss the questions in pairs and then take feedback as a class. Ensure that all the students understand the different land features and different types of water that are found on Earth. For question 5, students talk with a partner (or in small groups) about how they could find the answer to this question. Encourage them to refer back to their work from Unit 2.5.
- Read aloud the instructions on PCM ES5 and demonstrate how to make an 'ocean in a jar'. Make sure the students understand what to do before allowing them time to complete the activity in groups. Provide each group with a copy of PCM ES5, a clean glass jar with a secure lid, some water, blue food colouring, sand, shells and some other 'ocean' items. Ask them to make their own ocean in a jar. When they have finished, the students can try making waves by gently moving the jar from side to side.

Graded activities

1 Ask the students to make a model of the Earth. They should use a round balloon, blow it up and then cover it with layers of paper mâché. They can choose the size they wish to blow their balloon up to and how they want to paint their model, although they should attempt to show the difference between areas of land and water.

2 Ask the students to draw a picture of the Earth on Workbook page 80. Encourage them to add detail and colour to their picture. Ask them to label the areas of water and land. They should then complete the sentences using words from the unit. Ask questions to guide their thinking: *What covers most of the Earth's surface? Where can we find water on Earth?*

3 Remind students that much of the Earth's water is in the oceans and seas. Ask them if they can name any oceans or seas. Write some examples on the board. Next, ask the students to work in groups to research three oceans or seas of their choice. These can be taken from the examples on the board or the students can choose their own. Working in groups, help them to do some research using reference books or the internet. Students should make a poster to illustrate their findings. Display these on the wall in the classroom and discuss.

Consolidate and review

- Provide some globes for the students to look at. Ask them to identify their country on the globe. Write a list of other well-known countries on the board for them to locate.

Differentiation

■ All of the students should be able to make a model and use suitable colours, with little help.

● Most of the students should be able to draw and label an accurate representation of planet Earth and complete the sentences.

▲ Some of the students should be able to work collaboratively, thinking critically and asking questions to extend their knowledge and produce an informative poster. Some students may not be working at this level yet; if so, you may prefer to do this as a more structured activity by offering a frame for them to work to.

Earth and Space • Topic 6 Earth and Space

Science in context

6.3 Science and the environment

Student's Book pages 88–89

Earth and Space learning objectives

- *Planet Earth:* 1ESp.01 Know that Earth is mostly covered in water.
- *Earth in Space:* 1ESs.01 Know that Earth is the planet on which we live.

Science in context skills

- 1SIC.04 Talk about how science helps us understand our effect on the world around us.

Resources

- Workbook page 81
- Video ES2: Drought
- PCM ES6: Wasting water
- PCM ES7: Water saving game
- PCM ES8: Spinner
- PCM ES9: Game template
- PCM ES10: Spot the differences

Classroom equipment

- spinner for each group (made from PCM ES8)
- large sheets of paper, colour pens or pencils
- magazines to show examples of water uses

Key words

- Earth • water • waste • save

Scientific background

Water is a natural resource that is essential for all known life. Although water covers over 70 per cent of the Earth's surface, less than 3 per cent is fresh water. Fresh water is a precious resource, particularly in arid and semi-arid environments. Conserving water is important. Water resources in such regions are in short supply due to low rainfall and high evaporation rates. Droughts can happen when there is not enough rainfall over a prolonged period of time. Climate change is also beginning to cause environmental problems, such as reduced rainfall, which affects many countries and the world as a whole.

Water conservation is important in these areas and therefore methods of using water and ways to conserve it effectively must be implemented, to reduce wastage. There are many simple ways that we can all conserve water. These include turning the tap off while cleaning our teeth and washing our hands, only using the dishwasher and washing machine when they are full, and showering for a shorter time.

This unit gives students the opportunity to think about how science can help us to understand our effect on the world. It will hopefully encourage them to make small changes in their own everyday lives that can contribute towards helping the planet.

Introduction

- Ask students to look at the picture of the Earth on Student's Book page 88. Discuss the first question as a class. Recap that we live on the planet Earth and that most of the surface of the Earth is water. Remind the students that humans, animals and plants all need water to survive. (You may like to briefly recap Unit 2.5.)
- Show the class Video ES2 about a drought. Discuss what it shows. Tell the students that drought occurs when a place does not have enough water, usually in very hot countries. Explain to the class in simple terms that lack of rain can cause water shortages, which can destroy crops and cause hunger. Invite students to share their thoughts on this and discuss as a class. Establish that water is an important resource and we must not waste it, even in countries where droughts are less common. Tell the class that science can help us to understand our effect on the world and that in this unit they are going to look at ways to save water.

Teaching and learning activities

- Ask the students to look at the pictures on Student's Book pages 88–89. Allow them time to discuss the questions in groups. If necessary, prompt them by asking: *What can you see? Do you think there is enough water there? What do you think has happened to the plants? Do people need the plants to eat? Will they have enough food?* Take feedback as a class. Ensure that students understand that the dry environment has little water and the impacts this can have on people.

Earth and Space • Topic 6 Earth and Space

- Ask the class: *What do you use water for?* Discuss how students use water every day at school and at home. Write a list on the board. Reiterate the importance of water in our lives and stress that we should view it as a precious resource not to be wasted.
- Ask students to look at the picture on PCM ES6 and to circle the ways that water is being wasted (there are eight). They should then discuss how the people in the picture could have saved water.
- Ask the students to think again about how they use water at home. Remind them to think about all the different rooms and any areas outside, such as a garden or balcony. They should name some examples, such as cooking, drinking, flushing the toilet. Encourage the students to think of some less obvious examples, such as a washing machine or watering indoor plants. Ask them to write or draw pictures on Workbook page 81. Offer help as necessary. Ask: *How do we use water in the bathroom? Do we use water in the garden or for the car?* Then ask: *What can we do to use less water?* Allow the students time to discuss this in groups and ask them to make suggestions for how they can save water.

Graded activities

1 Tell the students that they are going to play a game about saving water. Give each pair a copy of PCM ES7 and a spinner that you have made from PCM ES8. Let the students spend some time playing the game. Circulate to ensure that they understand the concept behind the game: how water is used and how it can be saved.

2 Ask the students to review how they use water and ways they can use less water. Tell them that they are going to design a game about saving water. Give each group a copy of PCMs ES7 and ES9, and a spinner that you have made from PCM ES8. Ask them to devise their own version of the game using PCM ES7 for ideas. There should be penalties for wasting water and rewards for saving water. Ask the students to discuss the rules for playing their game. They can write the rules of their game on a separate piece of paper. Let the students play their game. Encourage them to concentrate on the issue on the importance of saving water. Ask if they had any problems playing the game: *Did the rules make sense? Was the game fair? Would you change anything about your game?*

3 Ask the students to design a poster to encourage people to save water, for display around the classroom or school as a reminder to others. Ask them to choose two or three ways we can save water. Help them to think of visual ways of showing this clearly. They can refer to the pictures in the Student's Book or look in magazines for ideas. Remind them that they do not need too many words, as pictures can explain a lot. Help them to come up with some simple sentences or labels to annotate their poster, such as: *Avoid wasting water! Turn off the taps when not in use!* Let the students present their work to the class and let them review each others' work. Ask: *Which posters do you like the best? Why? Are they easy to understand? Can we improve them? How?*

Consolidate and review

- Ask students to complete the spot the difference activity on PCM ES10.
- Ask students to discuss in groups changes they can make in their everyday lives to contribute towards helping the planet by conserving water.

Differentiation

■ All of the students should be able to work together as a group to play the game, taking turns and correctly following the instructions on the board.

● Most of the students should be able to identify and describe some obvious ways that we use and can save water, such as spending less time in the shower, and then work together to design a game using these ideas. For those who need some prompting, circulate, asking questions to guide their thinking.

▲ Some of the students should be able to identify and describe a wider range of ways to save water, including less obvious ones such as ensuring the dishwasher is full before using it. Offer support and guidance to students who may not be at this level yet.

Earth and Space • Topic 6 Earth and Space

6.4 What is land made of?

Student's Book pages 90–91

Earth and Space learning objective

- *Planet Earth:* 1ESp.02 Describe land as being made of rock and soil.

Thinking and working scientifically

- *Carrying out scientific enquiry:* 1TWSc.01 Sort and group objects, materials and living things based on observations of the similarities and differences between them; 1TWSc.05 Collect and record observations and/or measurements by annotating images and completing simple tables.

Resources

- Workbook page 82
- PCM ES11: Looking at soil

Classroom equipment

- a selection of different rocks and stones
- magnifying glasses
- sorting hoops
- a variety of soil types in trays
- modelling clay
- black paper, chalk pastels in a selection of colours
- small opaque bags

small plastic cups

Key words

- land • rock • soil

 Remind students to be careful with the stones and not to drop them or throw them. Make sure all the students wash their hands at the end of the lesson, as they have been handling soil.

Scientific background

The students learned in Unit 6.2 that approximately 70 per cent of the Earth's surface is water. The remaining 30 per cent is land, which is composed of rock and soil.

Rocks are large, solid objects that are made up of different minerals that occur naturally on the Earth's surface and in the outer solid layer of the Earth's crust. The natural weathering of rocks creates small pieces of rock, which may be called stones or pebbles. Pebbles are usually smoother than stones. They have been rolled and jostled together, either by water or wind, to create their smooth surfaces and rounded edges. Rocks are made of various minerals and in different ways, for example, through pressure or heat. They can be igneous, sedimentary or metamorphic.

Soil is a natural resource that is made from weathered rocks and detritus and is colonised by microorganisms. Soil is made up of sand, silt and clay, in different quantities, and contains solid particles, water and air. Soil is important in agriculture, mining and water management and is the most abundance ecosystem in the world.

Introduction

- Direct the students to look at the pictures on Student's Book pages 90–91. Ask them to discuss the questions in pairs. They should describe the two different landscapes and then think about what the stones and soil feel like. Ask questions to prompt them as necessary: *What does the rocky land look like? Can you describe the soil? What would it feel like to walk on the soil? What would it feel like to hold the stones/soil in your hands?* Take feedback as a class.

- Explain to the students that the Earth's land is made of rock and soil. Write the words *land*, *rock*, *soil* and *stone* on the board. Point out how each word is spelled and pronounced.

- Make sure the students understand the questions on Student's Book pages 90–91 and then discuss them as a class. This will serve as a useful recap of Unit 3.3 (properties). Ask students to name some other things that are made of rock. Encourage as many varied ideas as possible.

Teaching and learning activities

- Ask the students to look at the picture on Student's Book page 91. Ask what they think the children are doing. Tell them that they are going to do the same thing in this lesson: explore rocks and stones. They will also explore soil.

- Display a selection of interesting rocks and stones to discuss with the class. Ask: *Who likes this one?*

Earth and Space • Topic 6 Earth and Space

Why? What do you like about it? Allow the students time to look at the samples, to pick them up and to feel them. Demonstrate how to feel the rocks and how to use a magnifying glass to look at them more closely. Choose one with an interesting texture. Show it to the students and describe the surface using words such as *bumpy*, *smooth* and *rough*. Mention colour and shape, too. This will also serve as a useful review of Unit 3.3 (properties).

- Ask the students to choose their favourite rock or stone. Tell them to look closely at it and to draw a picture on Workbook page 82. Encourage them to draw it as accurately as possible. Ask questions, such as: *Is your stone all the same colour? Is it rough or smooth?* For activity 2, students study a small patch of land either at school or at home, and draw a picture of it.
- Place two hoops on the floor and give the students a selection of different rocks and stones. Ask them to sort them into the two hoops. First, ask them to separate the big stones from the small stones. When they have done this, ask them to sort by colour, for example *grey* and *not grey*. Ask the students to suggest other ways they can sort the stones.
- Explain briefly that soil is made of bits of broken rock and dead plants (old leaves), animal waste and air. Ask: *Why is soil important?* Clarify that we need soil because plants grow in soil. Without plants, humans and other animals would not survive.
- Give the students a variety of different soils in trays and some magnifying glasses. Ask them to look for differences in colour, texture and the size of the particles in the soils. Let the students feel the different soils with their hands, running their fingers through it and squeezing it. Discuss the different soil samples. Ask: *What colour is it? What does it feel like? Is it easy to squash*? Ask students to work in pairs to draw what they see through the magnifying glasses on PCM ES11, choosing the correct colours to match those that they see in the soil samples.
- Remind the class that there can be germs and bacteria in soil. Therefore, it is important that they always wash their hands after working with soil.

Graded activities

1 Give the students some modelling clay and a selection of stones and smaller pebbles of different colours, sizes, textures and patterns. Tell the students to roll out the modelling clay so that it is flat. They should then make a mosaic by pushing the stones and pebbles into the clay. Encourage them to use as many different textures, sizes and colours as they can.

2 Give each group of students a selection of stones and pebbles of different colours, textures and patterns, magnifying glasses, black paper and some coloured chalk. Ask the students to examine the stones carefully with the magnifying glasses and then draw some of the stones using the chalks. As they draw, encourage them to describe the stones using words such as *big*, *small*, *rough*, *smooth*, and to also describe their shape and colour. They should think carefully about the colour chalk that they use and how they can represent the texture of the stones on the paper.

3 Give each pair of students a selection of stones in small opaque bags, so that they can feel them but not see them. Let the students take turns to put a hand in the bag, feel a stone and describe it in as much detail as possible. They can then take the stone out of the bag to find out whether their description was accurate. Ask students which words they used to describe stones and make a class list on the board.

Consolidate and review

- Allow students some time to explore the soil samples again. Give each pair a small plastic cup so that they can make 'sand castles' by filling it with one of the soils and turning it out. Which soil kept the shape of the cup the best?

Differentiation

■ All of the students should be able to make an interesting mosaic using the clay and stones, with just a little help.

● Most of the students should be able to work together to create good representations of their chosen stones. Some may need support if they are unsure about how to show the texture of the stones.

▲ Some of the students should be able to use their sense of touch to select appropriate describing words for the stone that they feel, and then check visually to find out how accurate their description was. For those who need prompting, circulate, asking questions to guide their thinking.

Earth and Space • Topic 6 Earth and Space

6.5 The Sun

Student's Book pages 92–93
Earth and Space learning objective
- *Earth in Space:* 1ESs.02 Describe the Sun as a source of heat and light, and as one of many stars.

Thinking and working scientifically
- *Scientific enquiry: purpose and planning:* 1TWSp.01 Ask questions about the world around us and talk about how to find answers.
- *Carrying out scientific enquiry:* 1TWSc.05 Collect and record observations and/or measurements by annotating images and completing simple tables.

Resources
- Workbook pages 83 and 84
- PCM ES12: How bright?
- Slideshow ES2: The Sun

Classroom equipment
- paper plates, yellow paint, strips of red, orange and yellow tissue paper, glue

black paper, glue, glitter and chalk

Key words
- Sun • star • source • heat • light

 Supervise the students when they use glue. If you take the students on a walk around the school, ensure that they are safe and that they stay together. If the students use the internet, ensure they do so safely and always under adult supervision.

Scientific background

The *Sun* is a *star*. It is a gigantic ball of hydrogen at the centre of our Solar System. It looks much bigger and brighter than all the other stars in the sky because it is the nearest star to Earth.

The Sun is the *source* of all the Earth's energy. It produces *light* and *heat* as a result of nuclear fusion. Without heat from the Sun, the Earth would be a frozen and lifeless planet. All stars are light sources: they give out light, whereas planets are reflectors of light. Stars are always present in the sky but we only see them at night when the sky is not filled with light from the Sun. The Sun emits large amounts of light and is so bright it can easily damage your eyesight if you look at it directly.

The Earth, along with seven other planets, orbits the Sun. As the Earth orbits the Sun, it also spins on its axis and this is what gives us day and night. One full spin of the Earth takes one day. The half of the Earth facing towards the Sun experiences daytime and the opposite side experiences night-time. As the Sun goes down, the temperature drops, the sky changes colour and stars appear.

At this stage, the students only need to be able to describe the Sun as a source of heat and light and know that it is one of many stars in our sky.

Introduction

- Ask the class: *Who looked at the sky last night? What did you see? What colour was the sky? Did you see any stars?* Allow some discussion and write a list of class ideas on the board.

- Ask the students to look at the pictures on Student's Book pages 92–93. Explain that one of the images shows a close-up of the Sun. Introduce the fact that the *Sun* is a *star*, one of many (billions of) stars in the sky. Discuss the pictures and answer the first question as a class. Encourage the students to use the words *Sun*, *stars*, *blue sky* and *dark sky* in their responses.

- Ask: *When is the Sun in the sky? Is it hot or cool in the day? Is it hot or cool at night? What do you do in the day? Is it light or dark at night? What do you do at night?* Let the students discuss these ideas in groups. Elicit that when we can no longer see the Sun in the sky, it gets dark and cooler. Explain that this is because the Sun is a *source* of *heat* and *light*.

- Explain to students that the Sun is much closer to the Earth than other stars, which is why it looks bigger and we feel its effects (light and heat) more strongly. Tell the class that they are going to learn more about the Sun in this unit.

Earth and Space • Topic 6 Earth and Space

Teaching and learning activities

- Show Slideshow ES2, of the Sun and the Earth. Draw a picture of the Sun on the board. Ask: *What do we get from the Sun?* (heat, light) Ask students what they like and dislike about the Sun. Encourage them to share their ideas.
- Ask the students to look at the pictures on Student's Book pages 92–93 and to answer the questions in groups. Circulate, prompting as necessary: *What can you see in each picture? What time of day do you think it is? Why can we not see stars in the day? What would it be like if we did not have the Sun?*
- Ask the students: *What is the biggest light source you know?* Show the class some examples of light sources, such as a flashlight, a lamp and a picture of the Sun. Ask: *How are these things the same?* Tell them that they all give out light. Recap that the Sun is a star and that it is our biggest light source. Remind the class that looking at the Sun is very dangerous as it can damage our eyes and even blind us.
- Take students on a hunt to find as many different light sources as they can, inside and outside the classroom. They should record their findings on Workbook page 83 then choose one of the light sources to draw and label. Ask the students to share their findings with the class. How many different light sources did they find?

Graded activities

1 Show the students a paper plate that you have decorated as the Sun. Tell them they are now going to make their own Sun. Give each student the materials they need and demonstrate how to cut, tear and rip bits of coloured paper and stick them onto the edges of the plates once they have been painted. Display the Suns in the classroom. Ask: *What sense organ do we use to feel the heat of the Sun?* (the skin) *What sense do we use to see light from the Sun?* (sight)

2 Give each pair of students some cards cut from PCM ES12. Tell them to sort the pictures according to their order of brightness and stick them onto page 84 of their Workbook. Did all the students put the pictures into the same order. If not, why?

3 Ask students to design a fact sheet about the Sun. Elicit some ideas from the students about what they have learned in this lesson, helping them with any difficult vocabulary. Working in groups, help students to do some research on the Sun using reference books or the internet. They should include informative pictures and write clear annotations describing their facts about the Sun. Display the fact sheets on the wall and discuss. Ask each group to explain what they have learned about the Sun in this lesson. Ask: *What have you learned that is surprising? Interesting? Frightening?*

Consolidate and review

- Provide the students with some black paper, glue, glitter and chalk for them to make a starry night-sky scene.

Differentiation

■ All of the students should be able to decorate a plate to look like the Sun.

● Most of the students should be able to work together to correctly sort and then arrange the pictures into their order of brightness. Most should be able to explain their reasoning.

▲ Some of the students should be able to work collaboratively, thinking critically and asking sensible questions to extend their knowledge and produce an informative fact sheet. Some students may not be working at this level yet, so you may prefer to do this as a more structured task by offering a frame to work to.

Earth and Space • Topic 6 Earth and Space Consolidation

Consolidation

Student's Book page 94

Earth and Space learning objectives

- *Planet Earth:* 1ESp.01 Know that Earth is mostly covered in water; 1ESp.02 Describe land as being made of rock and soil.
- *Earth in Space:* 1ESs.01 Know that Earth is the planet on which we live; 1ESs.02 Describe the Sun as a source of heat and light, and as one of many stars.

Resources

- paper, colour pens or pencils
- Topic quiz sheets ES1 and ES2

Looking back

- Use the summary points on Student's Book page 94 to review the key things that the students have learned in the topic. Ask questions such as: *What is the name of the planet that we live on? What can you remember about the surface of the Earth? Can you name some water features that we can find on Earth? What is land made of? Can we use rock and soil for useful jobs? What is the Sun? What is the Sun a source of? Is the Sun important for life on Earth?* Encourage the students to use the key words from the topic in their answers.
- Ask the students to think about a place that they know well, and then to think what it looks like during the day and at night. Ask: *What is the same and what is different?* Remind them that the Sun is a source of heat and light, and encourage them to think about how this affects the place – light and warm during the day and dark and cooler at night. They should then draw and label pictures to show the place during the day and at night. Circulate to check that all the students understand what they need to do. This activity will show you how well the students have understood the topic.

How well do you remember?

You may use the revision and consolidation activities on Student's Book page 94 as a paired class activity. If you are using the activities to assess individual learning, have the students work on their own to complete the tasks in writing. If you are using them as a class activity, you may prefer to let the students do the tasks orally. Circulate as they discuss the questions and observe the students carefully, to see who is confident and who is unsure of the concepts.

Some suggested answers

1. Students' own drawings
2. Heat; light
3. Students' own answers
4. Students' own answers (cold and dark)

Consolidation

Consolidation and reinforcement of the students' understanding of the topic can be undertaken using Topic quiz sheets ES1 and ES2. These can be completed in class or as a homework task.

Topic quiz sheet answers

Sheet ES1

1. Students' labelled pictures
2. water; salty; lakes
3. true; false
4. rock; soil

Sheet ES2

1. ocean; lake
2. false; true
3. Earth; Sun
4. heat; light
5. The Sun should be circled.

Earth and Space • Topic 6 Earth and Space Student's Book answers

Student's Book answers

Pages 84–85 (6.1)
1 Taps – also bottles
2 We could get ill or possibly even die.
3 tap water; river water

Pages 86–87 (6.2)
1 Students' own answers (planet Earth)
2 blue = water; brown/green = land
3 The colour depends on the type of landscape.
4 Students' own answers
5 No, because it is salty.

Pages 88–89 (6.3)
1 Students' own answers (planet Earth/Pacific Ocean)
2 Students' own answers
3 Students' own descriptions
4 Students' own answers
5 'Slow the flow', i.e. turn off the taps
6 Turn the tap off.

Pages 90–91 (6.4)
1 Students' own descriptions
2 Students' own answers
3 wall; path
4 strong, hard, waterproof
5 Investigating/exploring rocks and stones

Pages 92–93 (6.5)
1 Students' own descriptions (stars shining in dark night sky; bright Sun in blue sky)
2 day (top image); night (bottom image)
3 Students' own answers
4 Because the light from the Sun is too bright during the day.
5 We need it for heat and light; without heat the planet would be too cold; without the Sun's light plants would not grow, humans and other animals would have no food, so we would not survive.

1.2 Is it alive?

PCM B1: Living or non-living?

Cut out the pictures and sort them into living and non-living things. Colour them in and use them to make a poster.

PCM B2: Plants or animals? (1)

Cut out the pictures and sort them into plants and animals.

1.3 Plants and animals are living things

PCM B2: Plants or animals? (2)

Cut out the pictures and sort them into plants and animals.

PCM B3: Five little leaves

Five green leaves, hanging on a tree,
Enjoying the sun, and swinging free,
Along came the wind, blowing past the shore,
Blew away a leaf, and then there were four.

Four green leaves, hanging on a tree,
Enjoying the sun, and swinging free,
Along came the wind, blowing from the sea,
Blew away a leaf, and then there were three.

Three green leaves, hanging on a tree,
Enjoying the sun, and swinging free,
Along came the wind, blowing through the bamboo,
Blew away a leaf, and then there were two.

Two green leaves, hanging on a tree,
Enjoying the sun, and swinging free,
Along came the wind, blowing from the sun,
Blew away a leaf, and then there was one.

One green leaf, hanging on a tree,
Enjoying the sun, and swinging free,
Along came the wind, its blowing almost done,
Blew away a leaf, and then there were none.

2.1 Parts of the human body

PCM B4: Where do you wear it?

Look at the pictures. Draw a line from each piece of clothing to the body part that you wear it on.

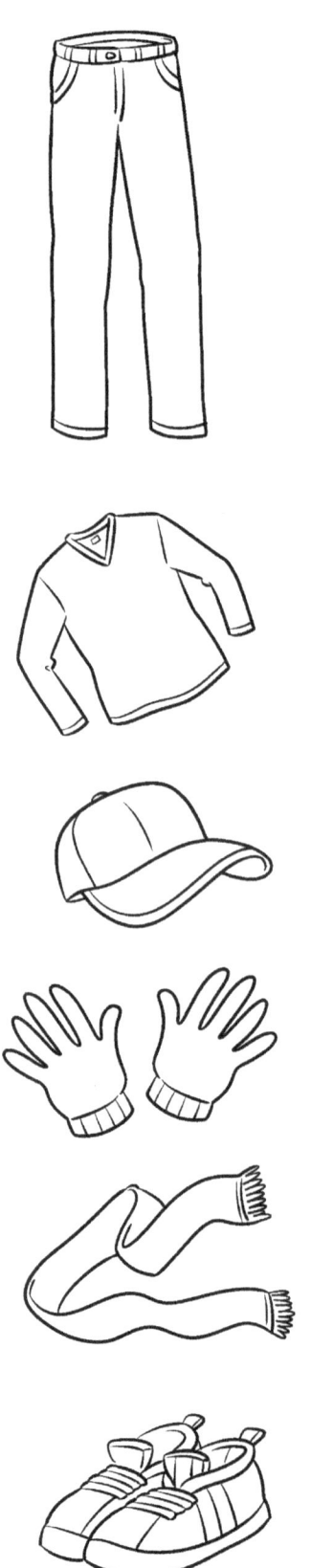
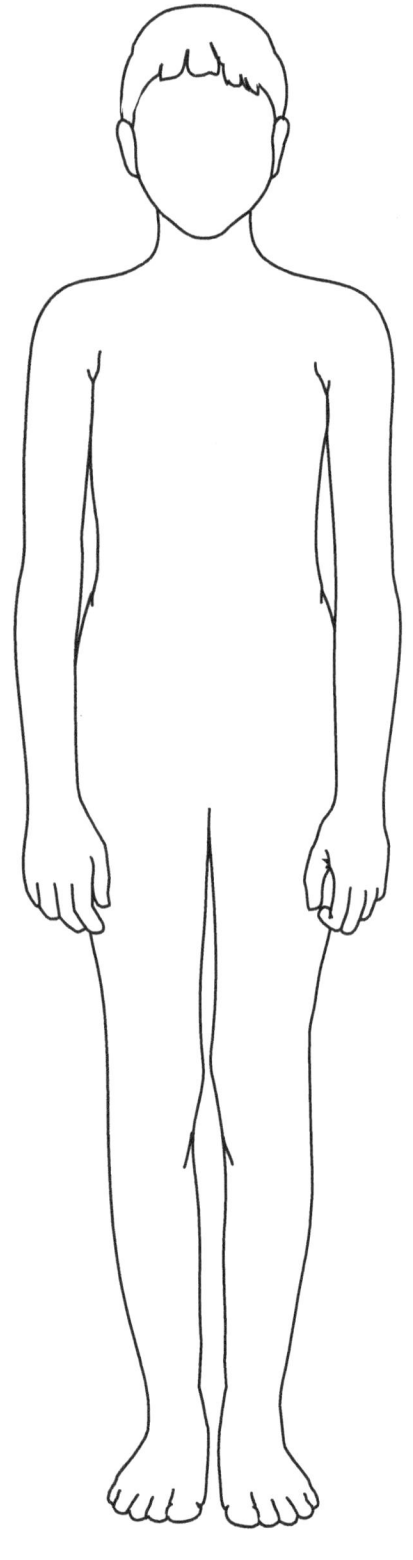

2.2 Our senses

PCM B5: Fruity smells

Smell each pot and draw a picture of the fruit that you think is inside the pot.

Drawing	Name
1	Were you correct?
2	Were you correct?
3	Were you correct?
4	Were you correct?

2.2 Our senses

PCM B6: Senses in action

What senses are the people in the picture using?

Person	Which senses are they using?	Why?

PCM B7: Whose eyes?

Draw a line from each eye to the animal the eye belongs to.

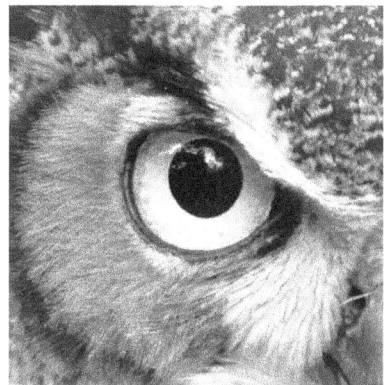
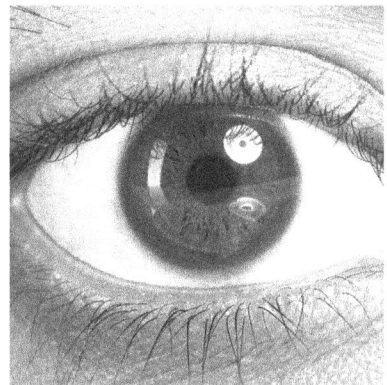
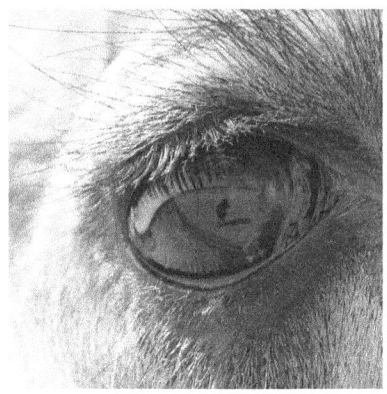

human

camel

frog

owl

2.5 What do animals need to survive?

PCM B8: Spot the differences

There are five differences to find. Draw a circle around each difference.

2.5 What do animals need to survive?

PCM B9: My family meals

Draw and label the different types of food that your family likes to eat.

My favourite food is _____

_____

Collins International Primary Science Stage 1 © HarperCollins*Publishers* 2021

2.5 What do animals need to survive?

PCM B10: **1–4 spinner**

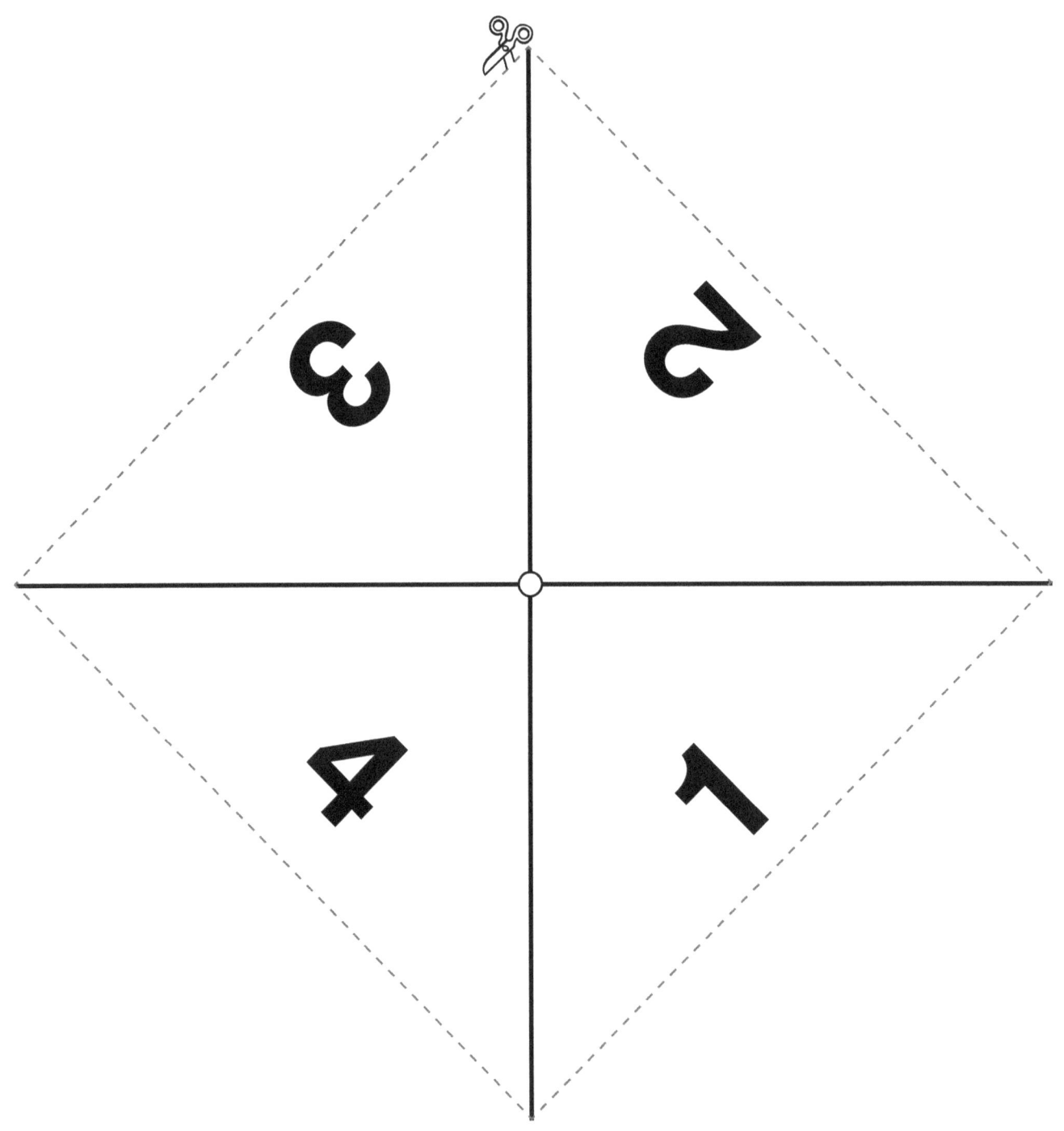

2.5 What do animals need to survive?

PCM B11: **Race to the water hole**

You are a gazelle trying to reach the water hole. Spin the spinner and move your counter along the board.

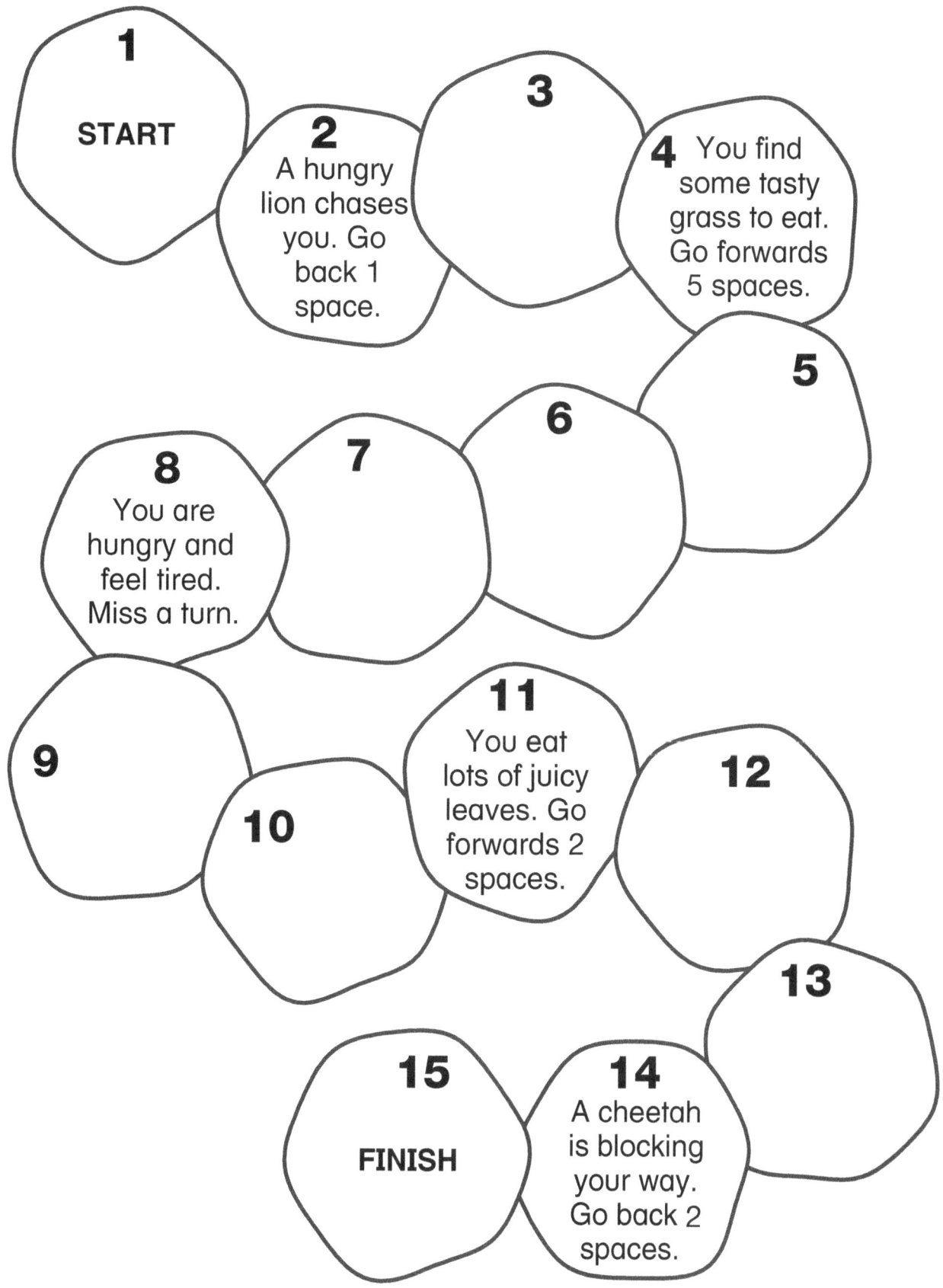

2.5 What do animals need to survive?

PCM B12: Animal spinner

2.6 Humans are similar

PCM B13: My family

Draw pictures of your family in the boxes below.
In what ways are your family's faces and bodies similar to yours?

This is _____

This is _____

This is _____

This is _____

2.7 Humans are different

PCM B14: Picture of my friend

Draw a picture of your friend in this box.

This is a picture of _____

I can tell who it is because of their

hair colour _____

hair style _____

eye colour _____

face shape _____

PCM B15: Spot the differences

There are five differences to find. Draw a circle around each difference.

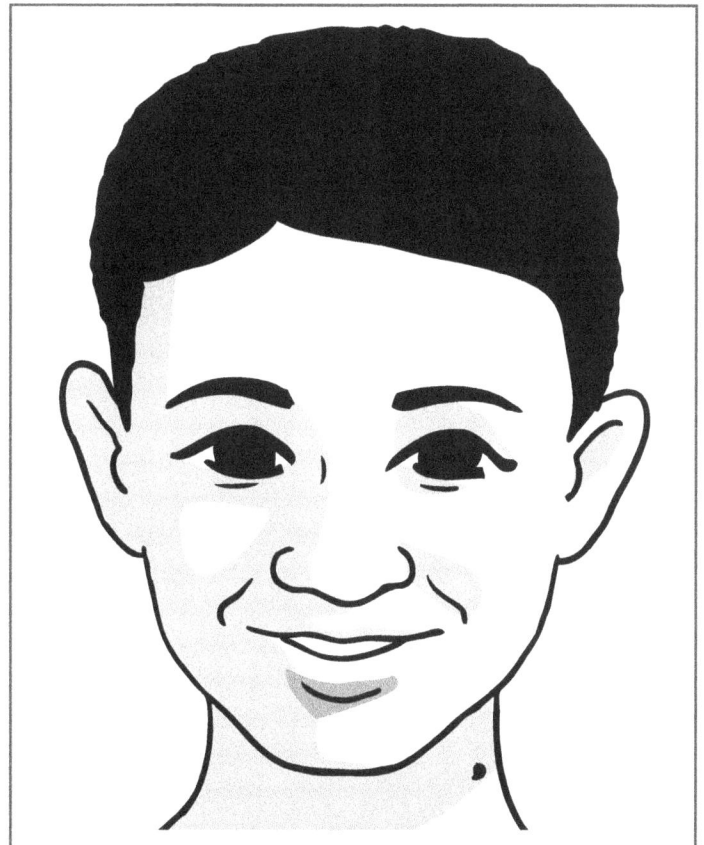

3.1 Similar or different?

PCM C1: Sorting into groups

PCM C2: Describing materials

3.3 More properties

PCM C3: Cardboard glasses

Copy on to card and cut out.

PCM C4: Concrete or glass?

List some objects made from concrete or glass that you might see on a tour of your school grounds.

Put a tick (✓) if you see the object. Put a cross (✗) if you do not see the object.

Objects made from concrete	Seen on tour?	Objects made from glass	Seen on tour?

PCM C5: **Sorting materials**

Sort the pictures into fabric, metal, wood and stone.

3.8 Materials can change shape

PCM C6: Elastic bands

You are going to test four elastic bands to see how much they stretch.

1 Which elastic band do you think will stretch the most? Why?

_____

2 Describe each elastic band. Then test each one. Measure how far each elastic band stretches. Record your results in this table.

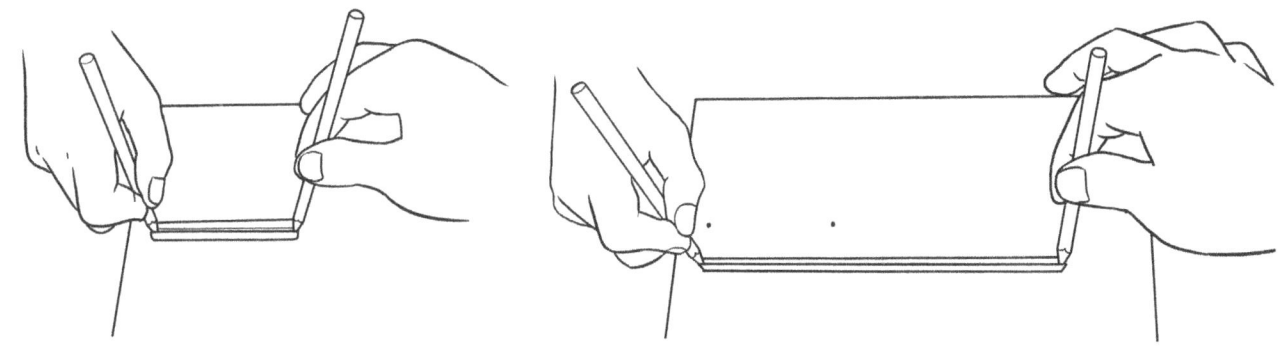

	Describe the elastic band	How far did the band stretch? (cm)
1		
2		
3		
4		

3 Which elastic band stretched the most?

_____

4 Look back at your prediction. Were you correct?

Yes ☐   No ☐   Not sure ☐

Collins International Primary Science Stage 1 © HarperCollins*Publishers* 2021

3.11 Uses of science

PCM C7: Which is the most waterproof?

Predict which material will be the most waterproof.
Why do you think this?

_____

_____

Diagram: a hand using a dropper to drop water into a yoghurt pot (with a hole in the bottom) containing material, sitting on a paper towel.

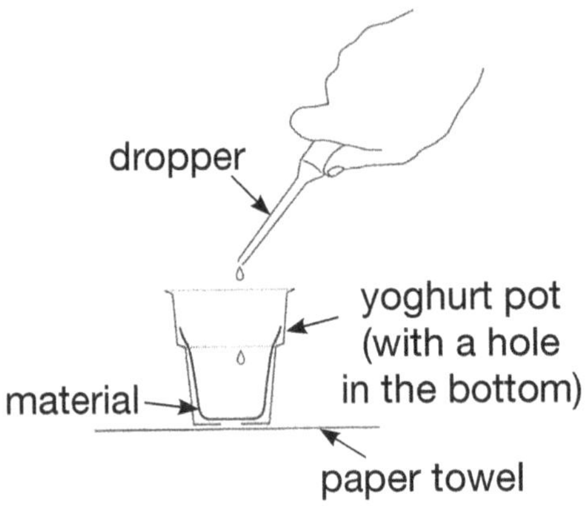

Test some materials.

Which material let the most water through onto the paper towel?

_____

Which material let the least water through onto the paper towel?

_____

Which material is the most waterproof? Why do you think this?

_____

_____

Was your prediction correct?

Yes ☐   No ☐   Not sure ☐

3.11 Uses of science

PCM C8: My favourite bridge

Research some bridges around the world.

Choose your favourite bridge and draw a picture in the box.

What materials is the bridge made from?

_____

_____

_____

What properties do the materials have?

_____

_____

_____

4.1 Thinking and working scientifically

PCM P1: How far does it go?

Roll a ball gently, using a small force. Then roll the ball more strongly, using a big force.

Draw pictures to show what happened.

This is what happened when I _____ the ball gently.

[   ]

This is what happened when I _____ the ball more strongly.

[   ]

PCM P2: Animal movement

PCM P3: Pushing and pulling

Draw toys that work by pushing and toys that work by pulling.

Toys that work by pushing	Toys that work by pulling

4.4 Pushes and pulls

PCM P4: Making a waterwheel

What you will need
- stiff cardboard
- a plate to draw round
- scissors and sticky tape
- skewer

1 Cut two large, matching circles from the cardboard.

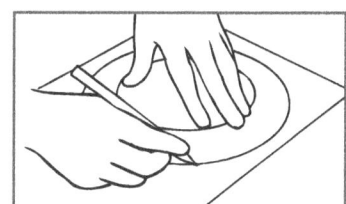

2 Cut a long strip of card, and then cut this into smaller, equally-sized, rectangular pieces.

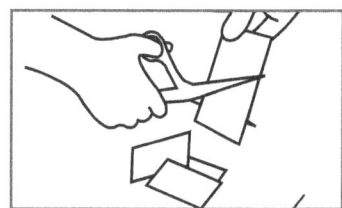

3 Take one circle of card and tape the rectangular pieces evenly around the edges.

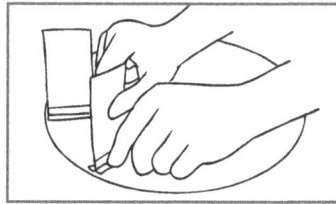

4 Now tape on the other cardboard circle.

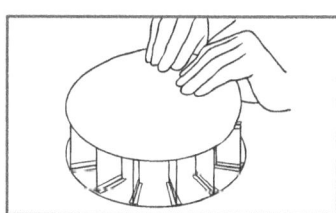

5 Put a skewer through the middle of the wheel, as shown.

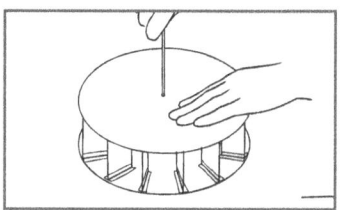

6 Hold both ends of the skewer.

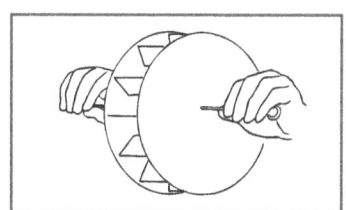

7 Hold the waterwheel under a gently running tap and watch it turn.

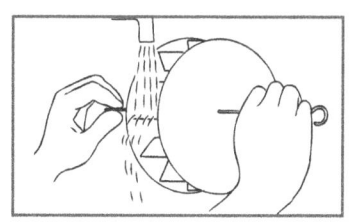

Collins International Primary Science Stage 1

4.5 Floating and sinking

PCM P5: Float or sink?

PCM P6: Will it float?

4.5 Floating and sinking

PCM P7: **Making paper boats** (sheet 1 of 3)

Boat 1

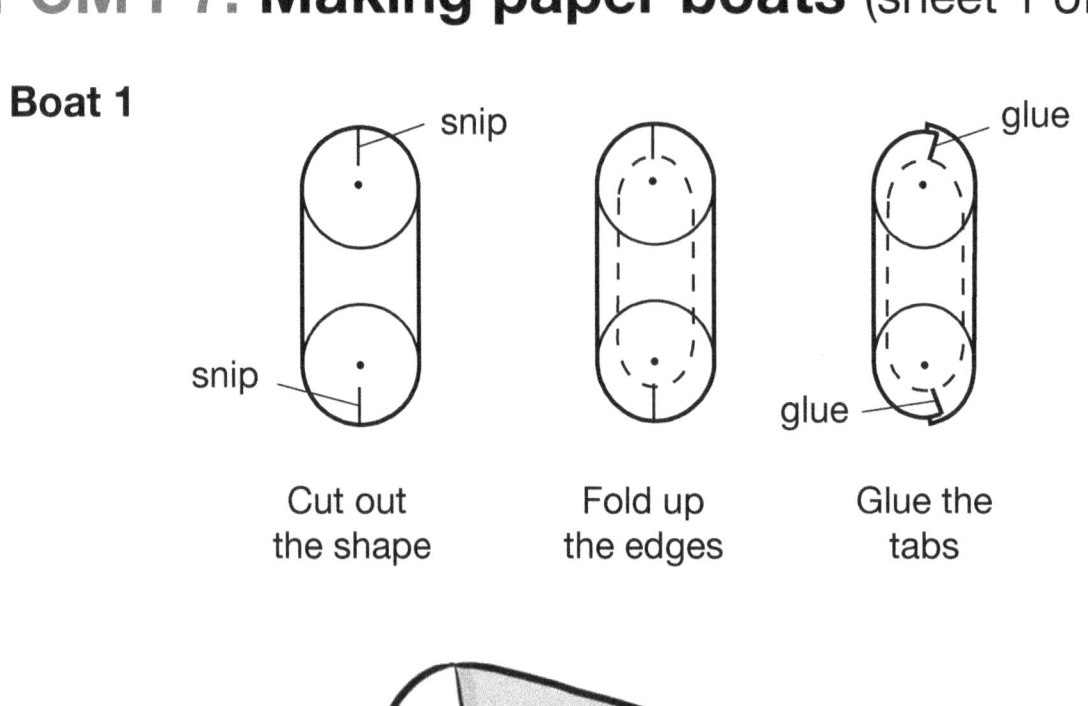

Cut out the shape

Fold up the edges

Glue the tabs

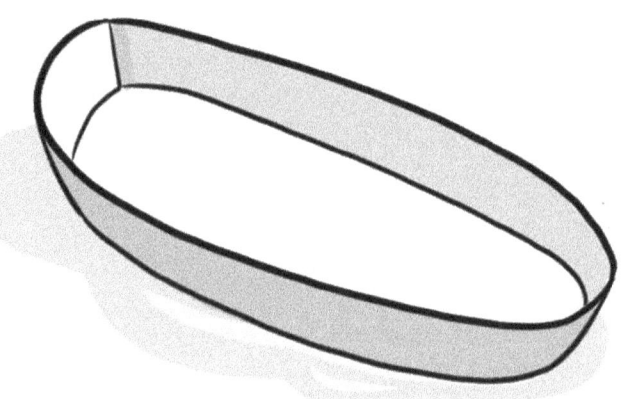

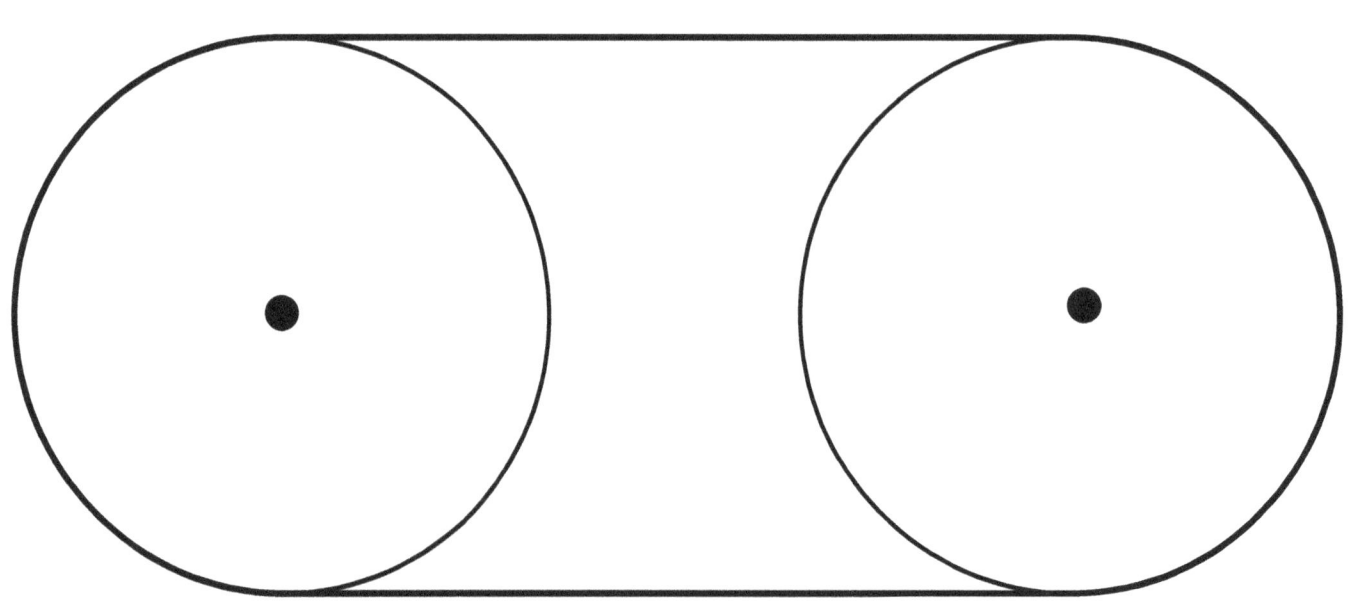

PCM P7: Making paper boats (sheet 2 of 3)

Boat 2

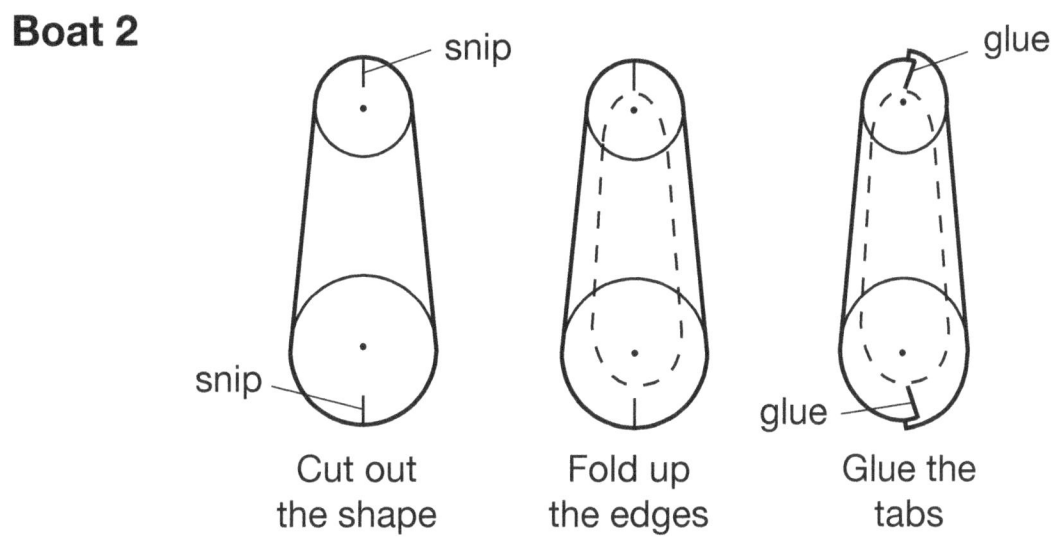

Cut out the shape | Fold up the edges | Glue the tabs

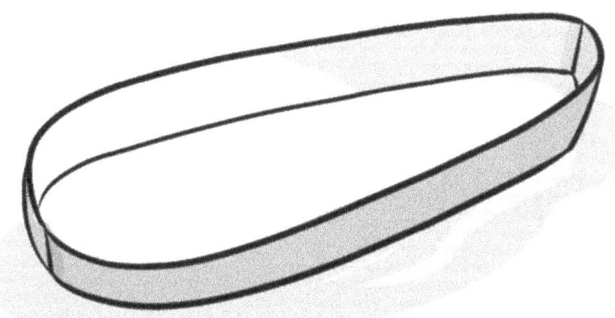

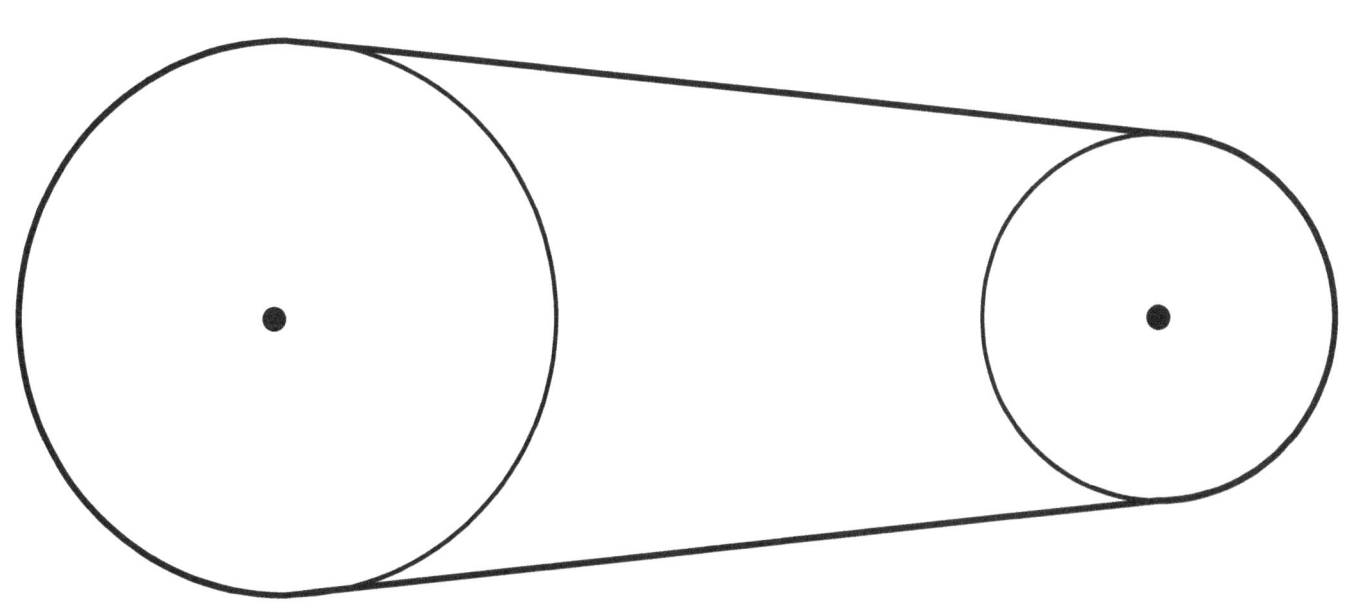

4.5 Floating and sinking

PCM P7: **Making paper boats** (sheet 3 of 3)

Boat 3

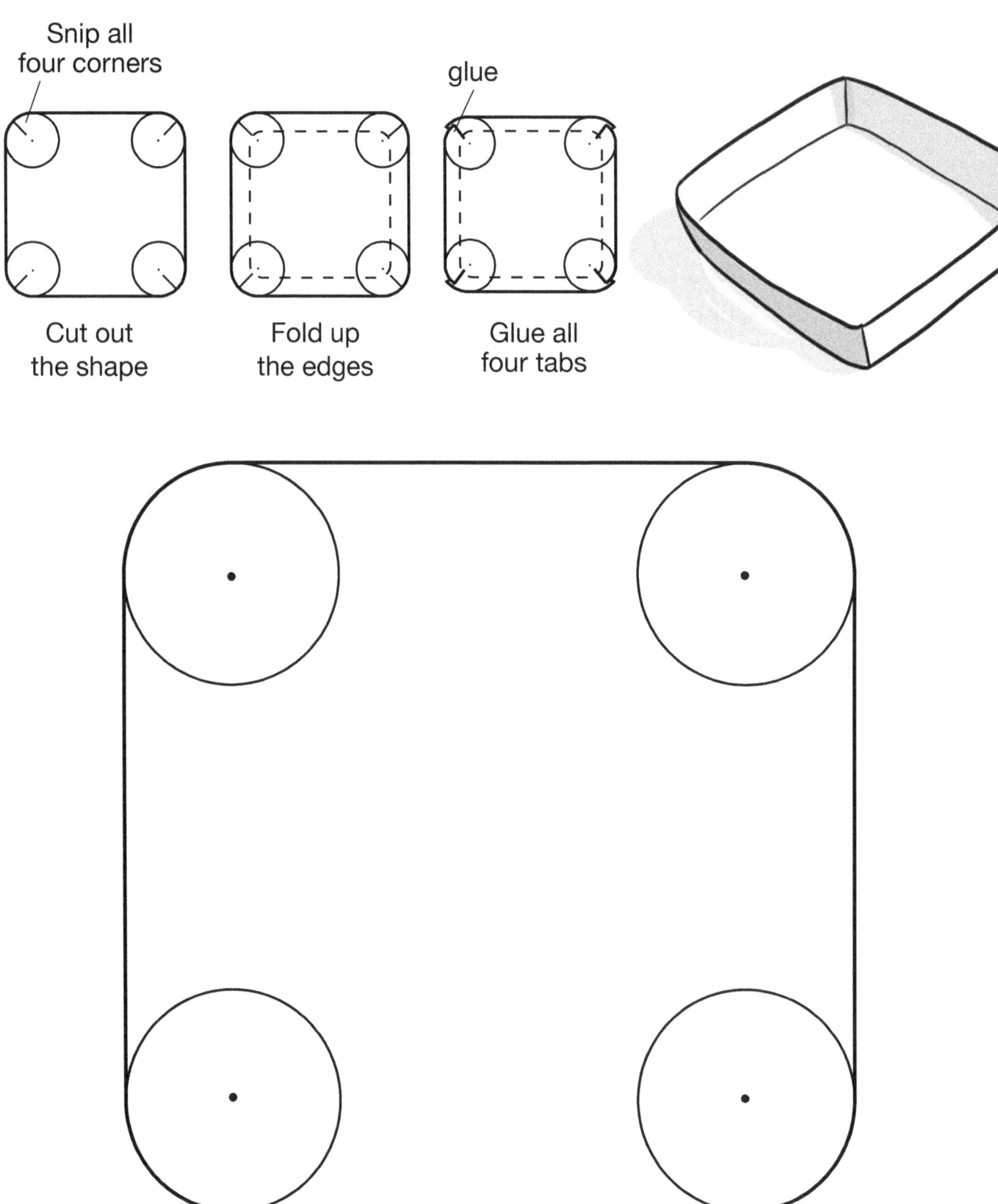

Snip all four corners

glue

Cut out the shape

Fold up the edges

Glue all four tabs

PCM P8: Sound source survey

Write or draw a picture of each sound you hear and its source.

Sound	Source
whistling	a bird in the tree

PCM P9: Musical instruments

Draw a line to show how to play each instrument.

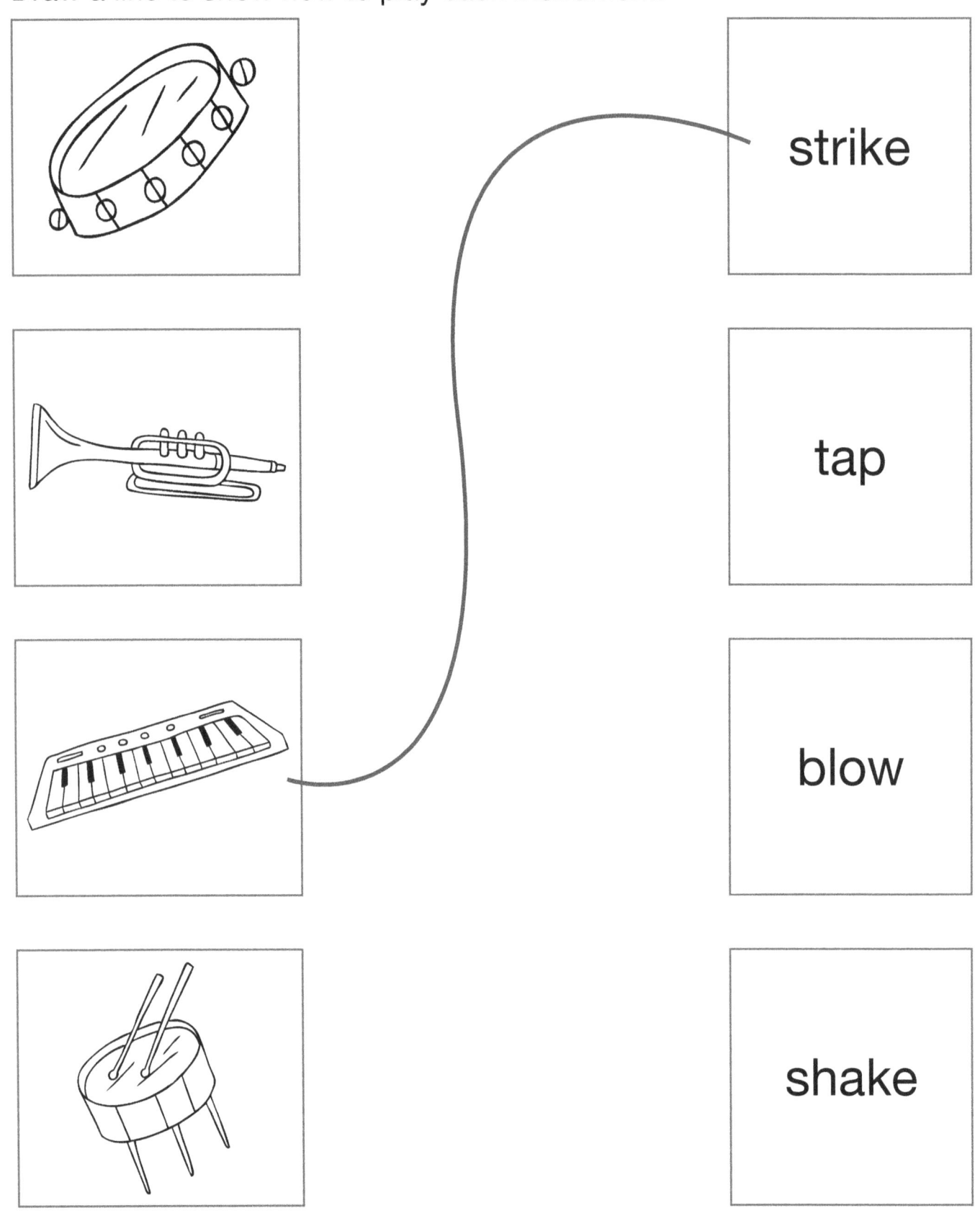

PCM P10: Sound word cards

strike	pluck
blow	shake

PCM P11: Blow, pluck or strike?

Draw an instrument in each circle that is played like this.

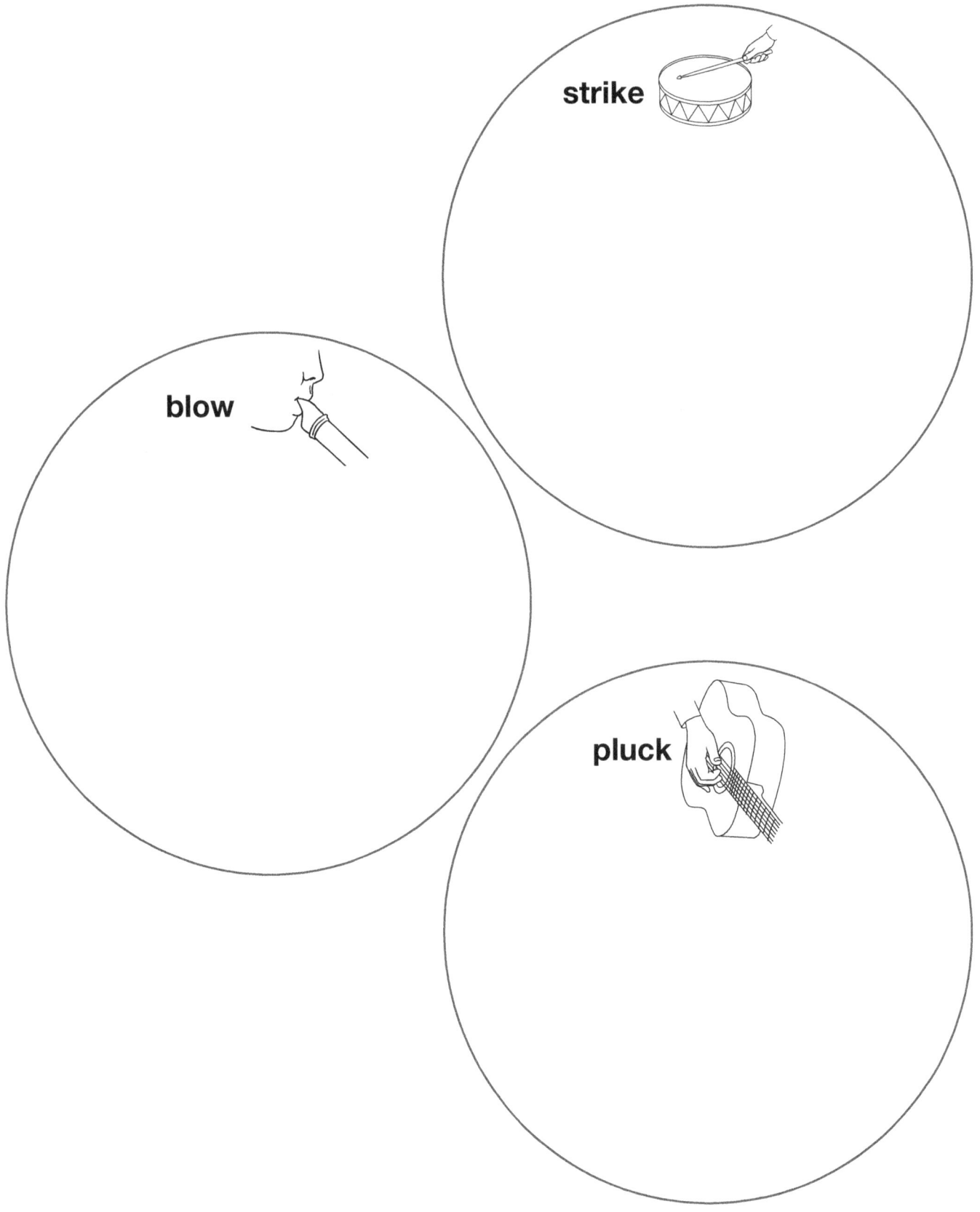

PCM P12: Making a shaker

What you will need
- empty plastic bottle with a lid
- items to put inside the bottle
- funnel

1

2

3
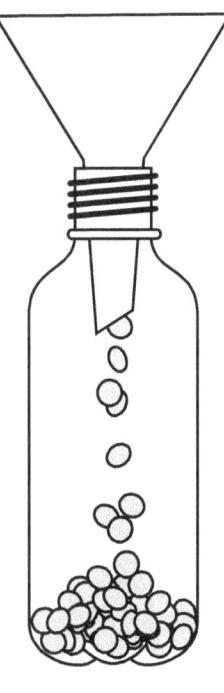

4

4.9 Sound and distance

PCM P13: Sound and distance

Investigate how far away you will be able to hear each sound.
Tick (✓) the correct answers.

Sound	Close by	A short distance away	A long distance away	Very far away
(bell)				
(pencil writing)				
(girls whispering)				
(aeroplane)				
(ambulance)				

5.1 What things need electricity?

PCM P14: Things that use electricity

Draw lines to match each device to what the device is used for.

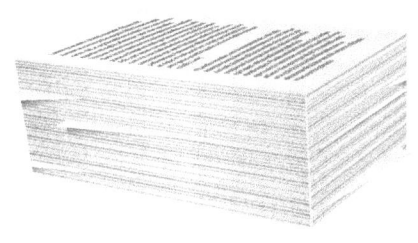

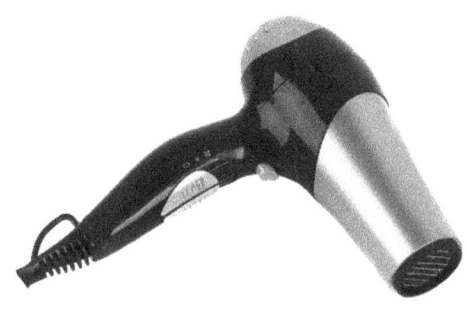

Collins International Primary Science Stage 1 © HarperCollinsPublishers 2021

PCM P15: Mains electricity or batteries?

Do these things use mains electricity or batteries?

Colour the objects that use mains electricity in red.

Colour the objects that use batteries in blue.

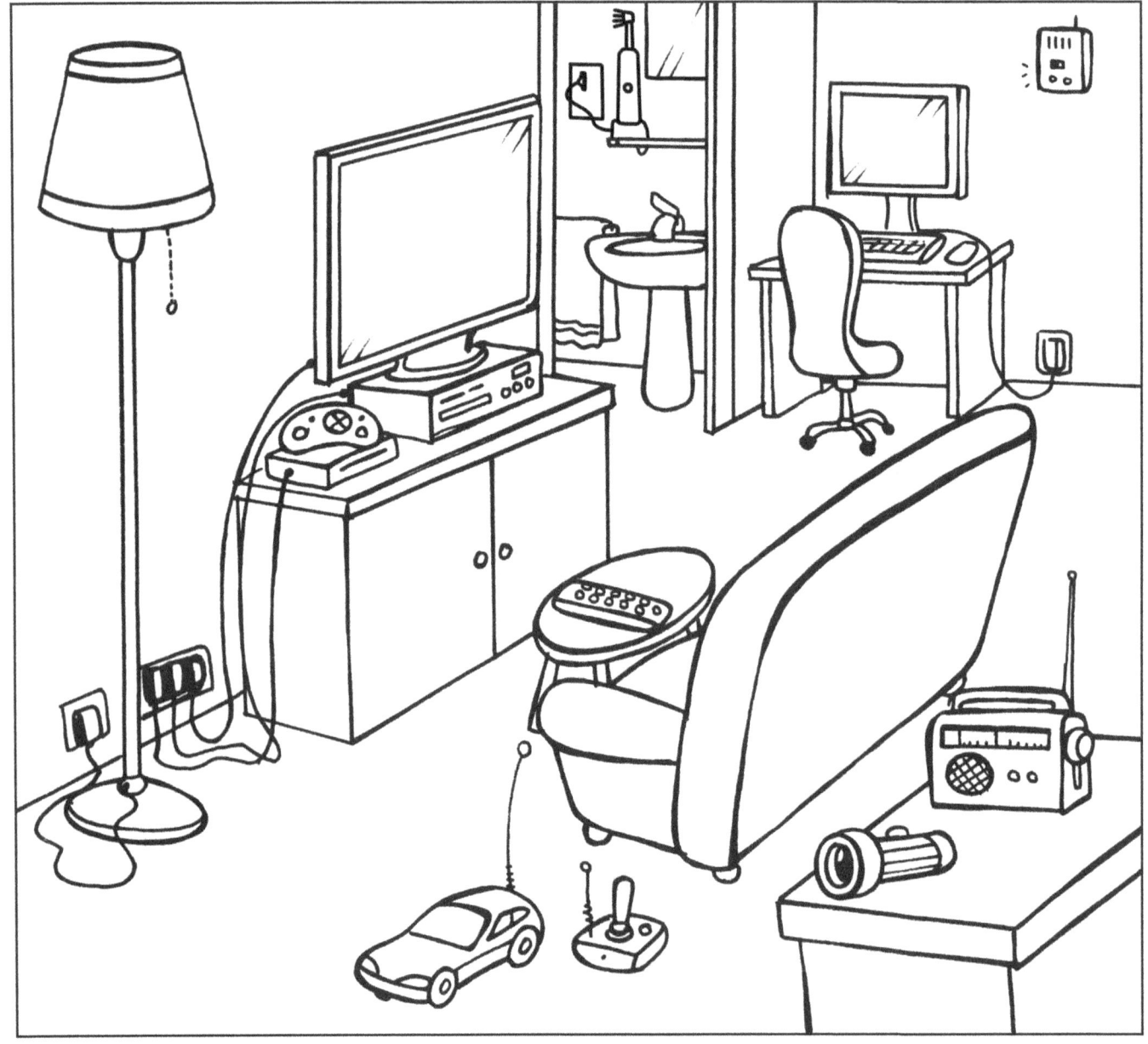

PCM P16: Magnetic or not?

Put a tick (✓) next to all the things that are magnetic.

PCM P17: No electricity

What could you do if there was no electricity?

I could use _____ for _____ .	I could use _____ for _____ .
I could use _____ for _____ .	I could use _____ for _____ .

What would you find hardest to do without electricity?

_____

_____

5.3 History of science

PCM P18: Then and now

 Early machines for adding numbers were big and very expensive.

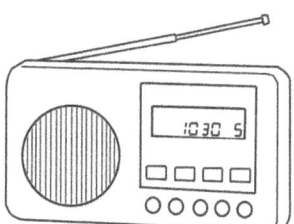 We can listen to music on a radio or computer.

 The first TV set had a very small screen with black and white pictures.

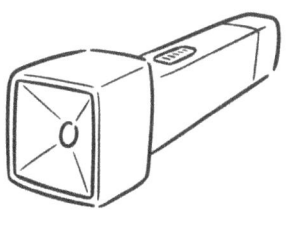 Modern battery flashlights are very bright and are light to carry.

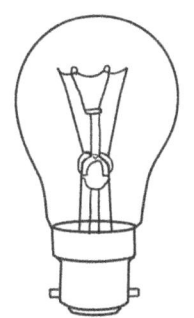 Old light bulbs gave out yellow light and wasted energy.

 A modern TV has a big screen and a colour picture.

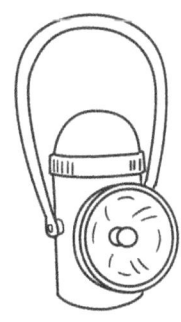 Old hand lanterns were heavy and gave out poor light.

 Modern light bulbs last a long time and do not waste much energy.

 People listened to the news and music on an old radio set.

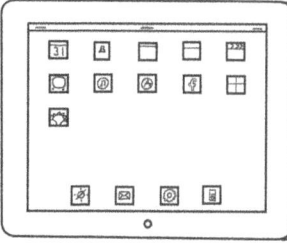 A modern tablet is a very fast computer that can do many different things.

Collins International Primary Science Stage 1

5.3 History of science

PCM P19: Before and after electricity

Choose two things that were different before and after electricity.
Draw and label pictures to show how they changed.

Before electricity	After electricity

Describe how they changed.

1 _____

_____

_____

2 _____

_____

_____

6.1 Clean water investigation

PCM ES1: Moving water

Predict which will be the best way to move water.
Predict which will be the worst.
Number them from 1 to 4, with 1 the best and 4 the worst.

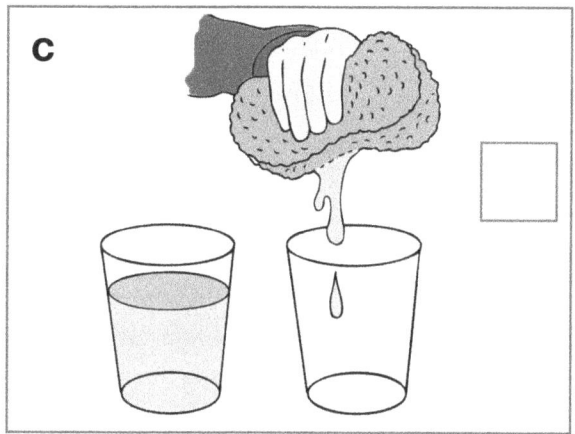
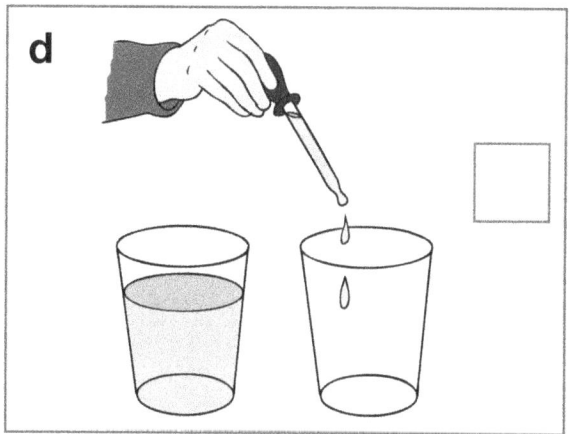

Describe what happened.

_____

_____

_____

_____

Were your predictions correct?

Yes ☐   No ☐   Not sure ☐

6.1 Clean water investigation

PCM ES2: Filtering water

What you will need
- 2-litre plastic bottle
- scissors
- cotton fabric
- elastic band
- cotton wool
- washed sand
- washed gravel
- water mixed with soil so it looks muddy

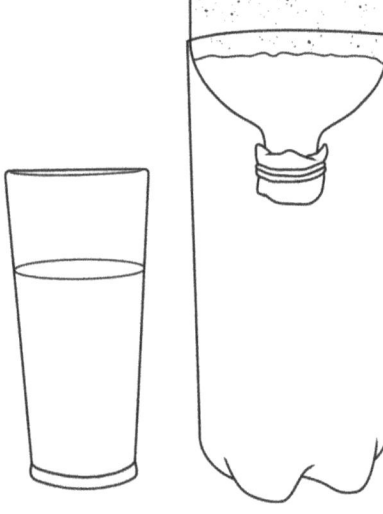

Introduction

Water from most sources must be cleaned before it is used because it contains impurities, bacteria and other micro-organisms that can make you ill.

This experiment demonstrates a simplified version of filtration.

Method

Cut off the top third of a plastic bottle (keep the remaining part).

Use an elastic band to secure a piece of cotton material over the spout of the bottle. Turn the top part of the bottle upside down and place it inside the bottom part of the bottle.

Add a layer of cotton wool to the bottom, followed by a layer of sand and then a layer of gravel.

Pour a glass of dirty water (water mixed with soil) through the filter.

Observations

The water looks much clearer once it has passed through the filter.

Explanation

The gravel filters the solid large particles that are not dissolved in the dirty water. The size of the particles that can be removed by filtration depends on the size of the filter you use. The gaps between the grains of sand are smaller than those between the pieces of gravel, so sand will stop more dirt particles from getting through than the gravel will. The cotton wool, and finally the cotton fabric, should remove even more.

Safety

The water must not be drunk even after filtering, as it may still contain bacteria from the soil.

PCM ES3: Testing water filters

Each group will need:
- 1-litre plastic drinks bottle cut in half, with the bottom part removed
- 10 cotton balls
- 5 paper napkins
- coffee filter
- plastic bag
- supply of dirty water for five investigations (add leaves, soil and sand to the water to make it dirty)
- beaker to catch the clean water in
- tray to carry out the investigation on

1. Turn the plastic bottle upside down.

2. Place the cotton balls into the neck of the bottle so it is sealed.

3. Now pour the dirty water into the bottle and catch what comes out in the beaker.

4. Draw a picture of the water that collects in the beaker.

5. Clean out the plastic bottle. Repeat with the other materials and see which makes the cleanest water.

Although the water appears clean, it is not safe to drink.

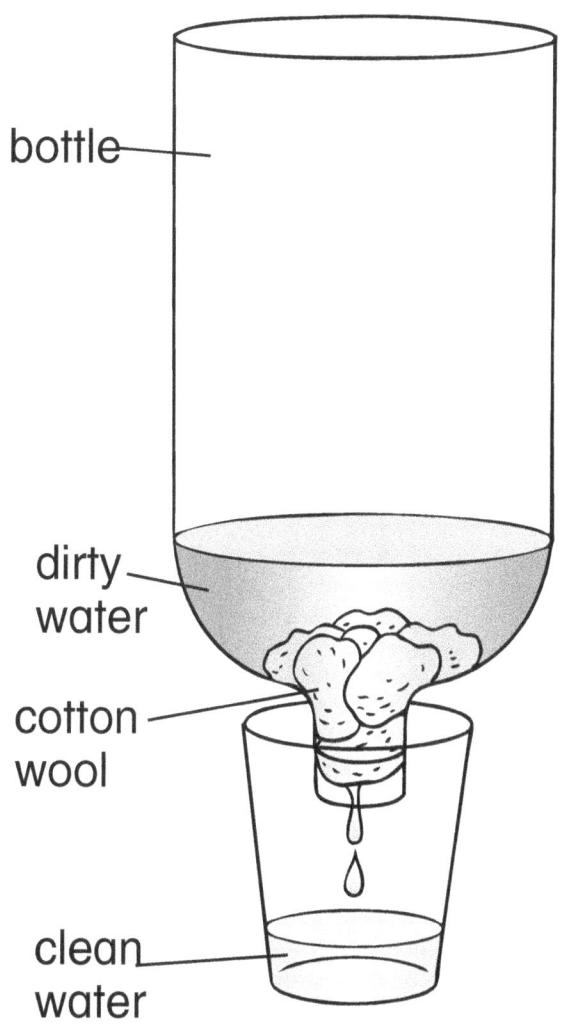

6.2 Our planet Earth

PCM ES4: Water on Earth

Draw and label one type of water that we can find on the Earth.

PCM ES5: **Ocean in a jar**

> **You will need:**
> - a glass jar
> - sand
> - shells, rocks or other ocean objects
> - water
> - blue food colouring

Instructions

1 Add a layer of sand to the bottom of the jar.

2 Add shells, rocks or other ocean objects.

3 Add water so that the jar is three-quarters full.

4 Add a few drops of blue food colouring to make the 'ocean' blue.

5 Place the lid on the jar and secure it tightly.

6 Gently rock the jar so you can see waves in the 'ocean'.

6.3 Science and the environment

PCM ES6: Wasting water

Draw a circle around each place where water is being wasted.

6.3 Science and the environment

PCM ES7: Water saving game

Play in pairs. Spin your spinner and move your counter. The first person to reach the finish is the winner.

Start

1

2 left tap running while cleaning teeth
Go back 2

3

4 cut shower time by 1 minute
Go forward 2

5 used bathroom towels more than once
Go forward 1

6

7 collected rainwater to water indoor plants
Go forward 4

8

9 only filled dishwasher half full
Go back 3

10 fixed a dripping tap
Go forward 1

11

12 washed car with a hosepipe
Go back 4

Finish

Collins International Primary Science Stage 1

6.3 Science and the environment

PCM ES8: Spinner

6.3 Science and the environment

PCM ES9: Game template

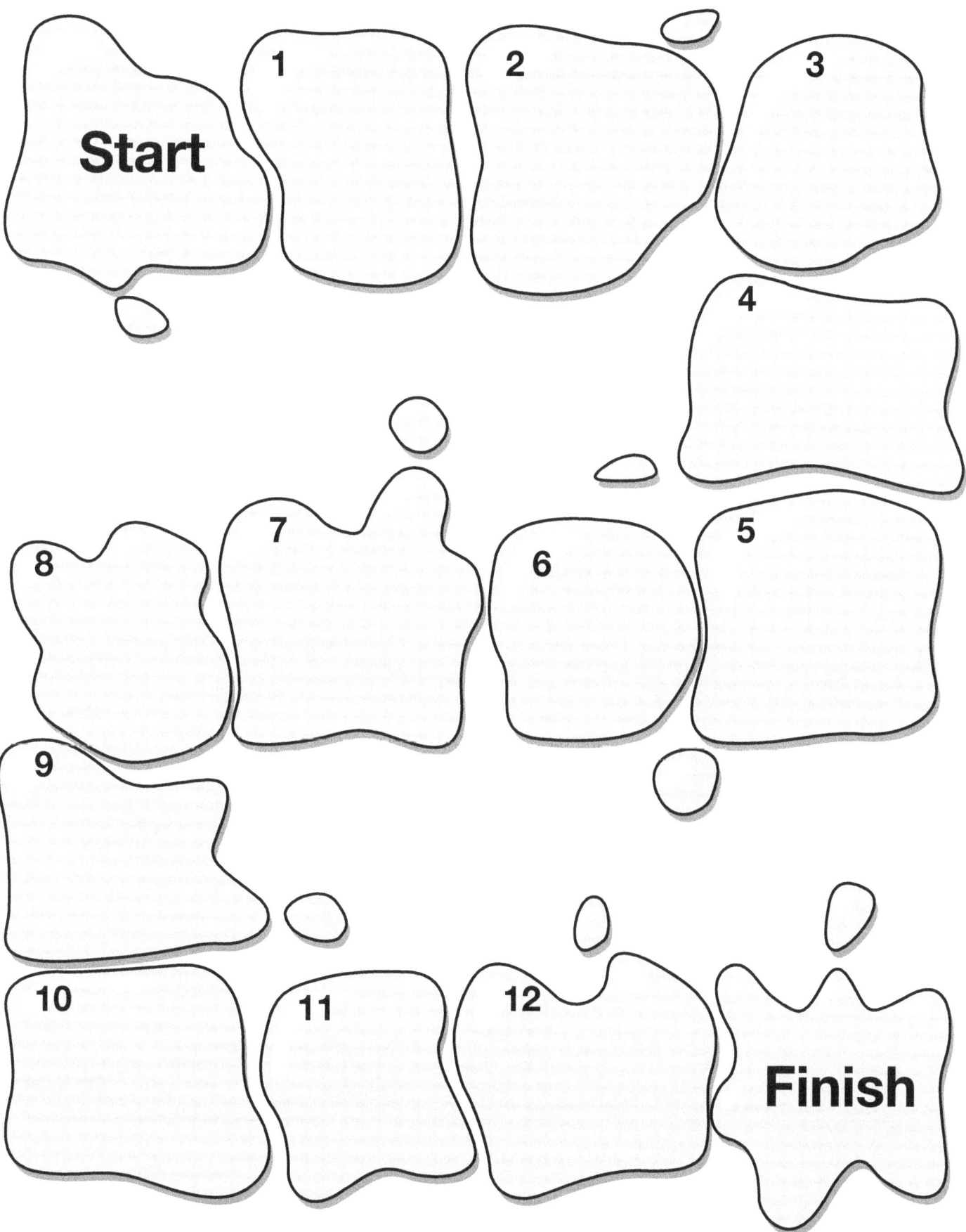

6.3 Science and the environment

PCM ES10: **Spot the differences**

Circle the five differences in picture B.

6.4 What is land made of?

PCM ES11: Looking at soil

Draw your soil pictures here.

6.5 The Sun

PCM ES12: How bright?

flashlight		lamp	
the Moon		window	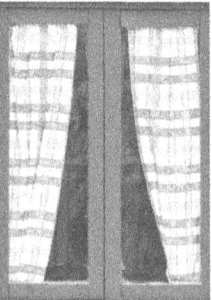
candle		mirror	
led display		glitter ball	

Topic 1 Plants

Biology: Topic quiz sheet B1

1 Draw three living things.

2 Draw three non-living things.

3 Which thing is the odd one out? Tick (✓) the correct box.

   tree ☐   elephant ☐   football ☐

   Why is it the odd one out?

   _____

4 True or false? Circle the correct answer.

   A wooden chair is a living thing.   True   False

5 Name one thing that has never been alive.

   _____

The Topic quiz sheets on pages 153–166 have been written by the authors. These and references to assessment and/or assessment preparation are the publisher's interpretation of the curriculum framework requirements and may not fully reflect the approach of Cambridge Assessment International Education.

Topic 1 Plants

Biology: Topic quiz sheet B2

1. Label the picture of the plant. Use the words in the box to help you.

| stem   flower   leaf   root |

2. Which two roots can we eat? Circle the correct answers.

   coconut   sweet potato   tree   carrot   sunflower

3. Which plant is the odd one out? Tick the correct box.

   Why is the plant the odd one out? _____

4. True or false? Circle the correct answers.

   All plants need water.   True False

   Plants can live without light.   True False

Topic 2 Humans and other animals

Biology: Topic quiz sheet B3

1 Label the parts of the body. Use the words in the box to help you.

| head finger foot arm chest hand neck leg |

2 What features do a human, a frog and a horse have in common?

_____

Topic 2 Humans and other animals

Biology: Topic quiz sheet B4

1 Draw a line from each animal to its food.

2 Circle the correct word in each sentence.

All animals need **food** / **light** to survive.

Seawater has **salt** / **sugar** in it and is not good to drink.

3 True or false? Circle the correct answers.

All animals eat plants.   True   False

Animals cannot live without food and water.   True   False

Dirty water is safe to drink.   True   False

Topic 2 Humans and other animals

Biology: Topic quiz sheet B5

1. Look at the picture of the boy. Draw a line from each sense word to the correct sense organ.

sight

smell

touch

hearing

2. Look at the pictures of the boys.

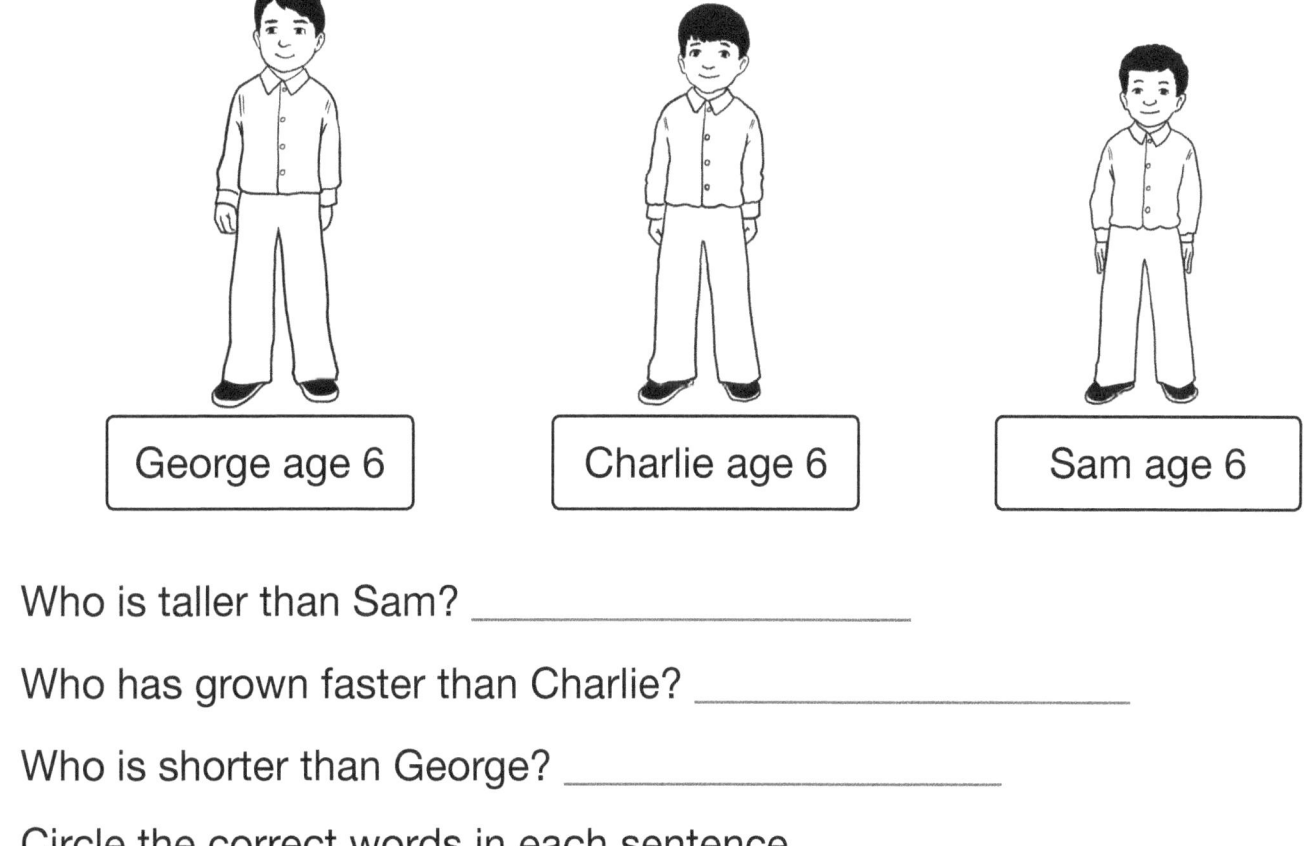

George age 6

Charlie age 6

Sam age 6

Who is taller than Sam? _____

Who has grown faster than Charlie? _____

Who is shorter than George? _____

3. Circle the correct words in each sentence.

When you cross the road, you should **look** / **smell** and **touch** / **listen** to stay safe.

Our senses keep us **safe** / **sad**.

Topic 3 Materials

Chemistry: Topic quiz sheet C1

1. Name something that can be made from each material. Use a word from the box to describe one property of the material.

Material	What can be made from it?	Property of the material
plastic		
stone		
wood		
glass		
paper		

shiny	heavy	strong	smooth	hard
light	rough	flexible		transparent
soft	dull	waterproof		absorbent

Topic 3 Materials

Chemistry: Topic quiz sheet C2

1 Name two hard materials.

   _____

2 Name two soft materials.

   _____

3 Circle the absorbent materials.

4 Tick (✓) three properties of glass.

   transparent ☐

   smooth ☐

   elastic ☐

   easy to clean ☐

   absorbent ☐

5 True or false? Circle the correct answer.

   Concrete buildings are very strong.   True   False

Topic 4 Forces and sound

Physics: Topic quiz sheet P1

1   Draw a line from each animal to the way the animal moves.

swims

walks

flies

slithers

2   Circle each picture that shows a pull.

3   Circle the correct word in each sentence.

A sailing boat moves when the wind **pushes** / **pulls** its sail.

A bigger pushing force will make the boat move **slower** / **faster**.

4   Complete the sentence using words from the box.

Water is used to turn a waterwheel. Water can _____ with a very strong _____.

| push | pull | force | sound |

Topic 4 Forces and sound

Physics: Topic quiz sheet P2

1 Circle the toy that moves by pulling.

2 Three toy cars were pushed along a track.

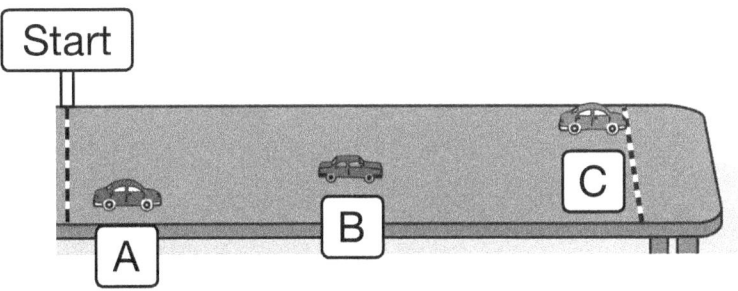

Which car, A, B or C, was given the biggest push? _____

Which car was given the smallest push? _____

Would a toy car move if it was not pushed? _____

3 True or false? Circle the correct answer.

Heavy objects are easy to push and pull.   True   False

Living things do not move.   True   False

4 Name two parts of your body that move when you are playing football.

_____

5 Tick (✓) the correct boxes.

What will make a waterwheel turn faster?

more water ☐   warmer water ☐   faster water ☐

Topic 4 Forces and sound

Physics: Topic quiz sheet P3

1 Tick (✓) the things that are sources of sound.

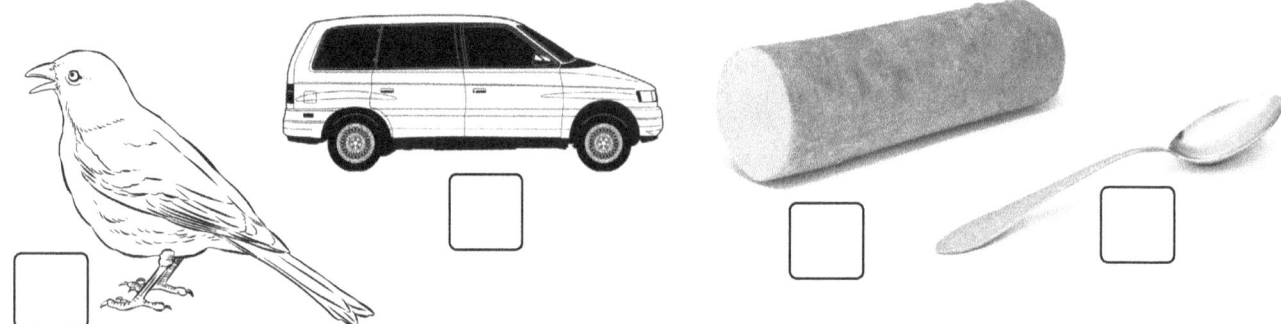

2 Draw three sources of sounds that you can hear in classroom.

3 Which sense organ do we use to hear with? Circle the correct answer.

ear   eye   nose   mouth

4 Name two natural sources of sound.

_____

5 Name two human-made sources of sound.

Physics: Topic quiz sheet P4

1 Draw a line to match each instrument to the way the instrument is played.

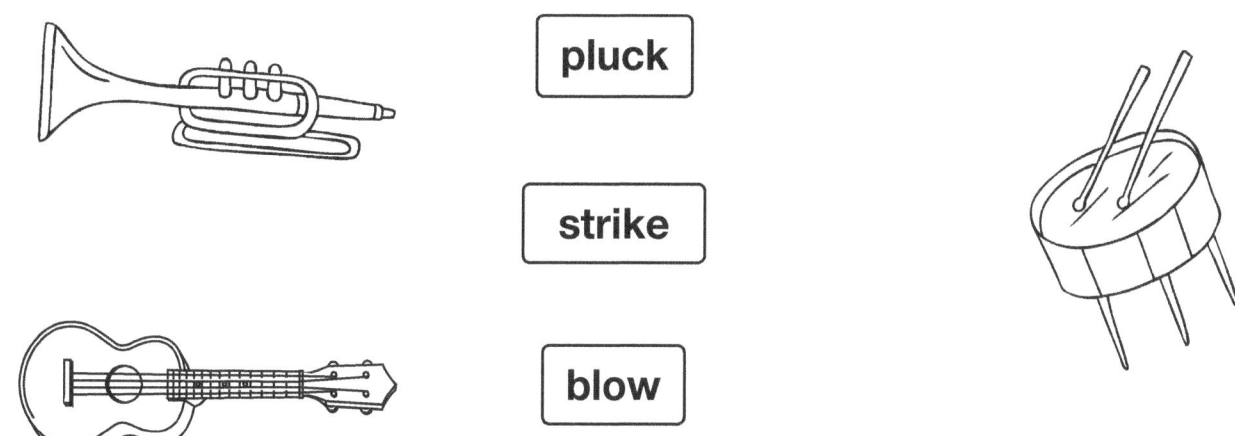

2 Circle the correct word in each sentence.

The sound from a source that is far away is **quiet** / **loud**.

To make a quiet clapping sound, I must clap my hands very **hard** / **gently**.

3 What two things can you do to protect your hearing from loud sounds? Tick the boxes.

move further away from the source ☐

close your eyes ☐

wear ear defenders ☐

4 Which of these are warning sounds? Circle them.

a fire alarm   a police siren

a cat purring   a telephone ringing

5 True or false? Circle the correct answer.

The wind is a natural source of sound.   True   False

Topic 5 Electricity and magnetism

Physics: Topic quiz sheet P5

1. Which of these devices use electricity?
   Draw a line from each device to the word.

   electricity

   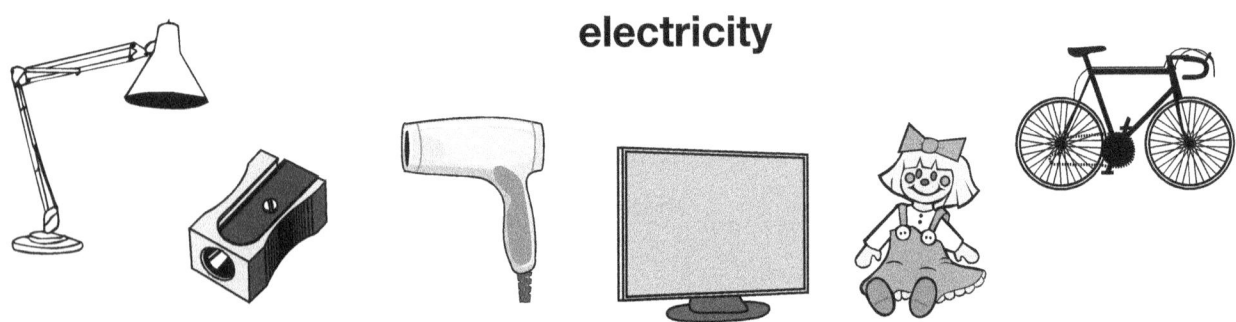

2. Name two devices that use a battery.

   _____   _____

3. Circle the magnetic objects.

   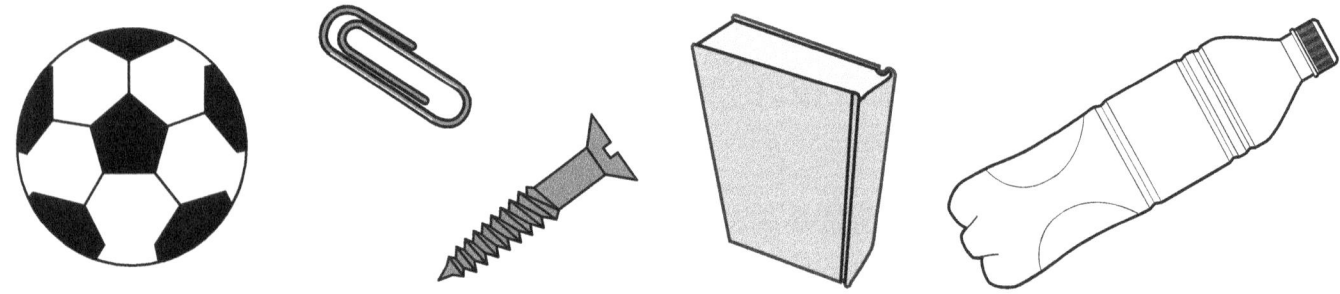

4. Complete the sentences using words from the box.

   | pull   metal   towards |

   A magnet is a special type of _____.

   Magnetic materials move _____ a magnet.

   If you hold a magnetic material near a magnet you

   feel a _____.

Earth and Space: Topic quiz sheet ES1

1 Label this picture using words from the box.

| land   water   Earth |

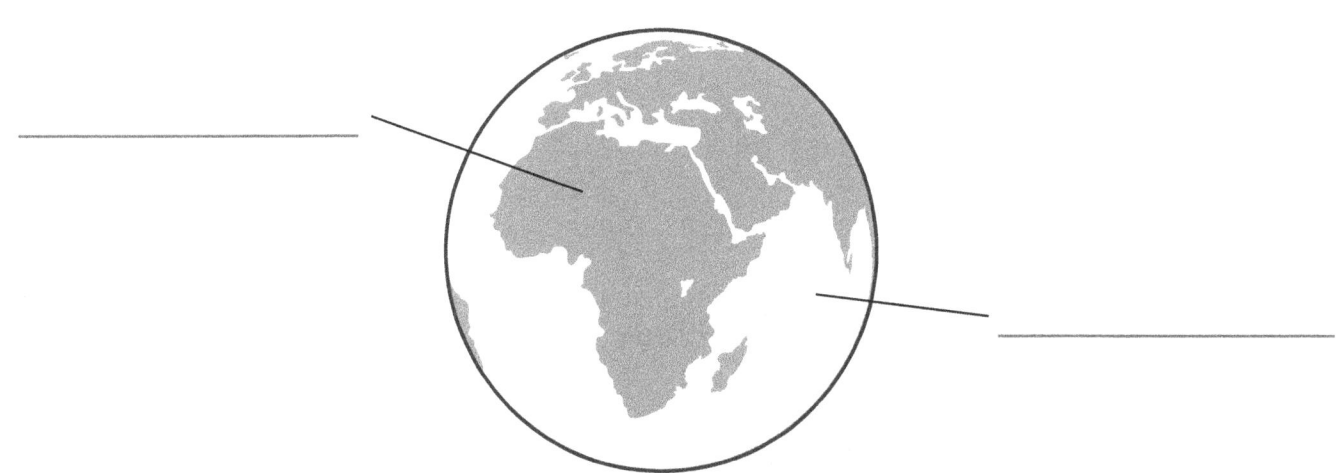

Our planet _____

2 Circle the correct word in each sentence.

Most of the Earth's surface is covered by **land** / **water**.

The water in the oceans is **fresh** / **salty** water.

Lakes / **Forests** are a type of water feature on Earth.

3 True or false? Circle the correct answer.

The Sun is a source of heat.   True   False

The Earth is a source of light.   True   False

4 Name the two things that land is made of.

1 _____

2 _____

Earth and Space: Topic quiz sheet ES2

1 Tick (✓) the water features that we can find on Earth.

desert ☐   ocean ☐   mountain ☐   lake ☐

2 True or false? Circle the correct answer.

The Earth is a star.   True   False

The Earth's biggest source of light is the Sun.   True   False

3 Complete the sentences using words from the box.

Sun   Earth

The planet that we live on is called _____.

The _____ is a large star.

4 Circle the things that we get from the Sun.

heat   water   light   food

5 Circle the brightest light.

   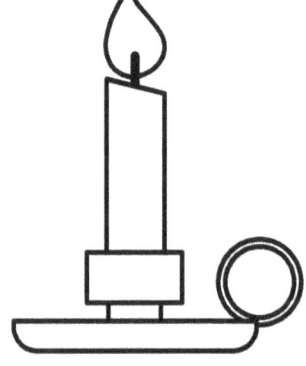   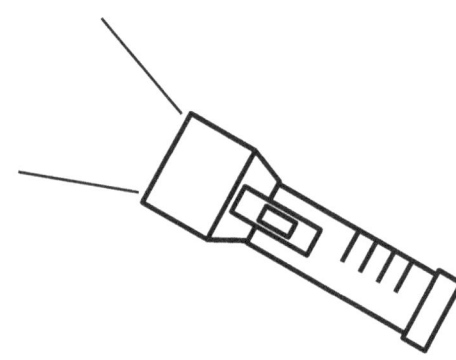

Cambridge Global Perspectives™

Below are some examples of lessons in *Collins International Primary Science Stage 1* which could be used to develop the Global Perspectives skills. The notes in *italics* suggest how the Science activity can be made more relevant to Global Perspectives.

Please note that the examples below link specifically to the learning objectives in the Global Perspectives curriculum framework for Stage 1. However, skills development in a wider sense is embedded throughout this course and teachers are encouraged to promote research, analysis, evaluation, reflection, collaboration and communication as general best practice.

Cambridge Global Perspectives	Learning Objectives for Stage 1	Collins International Primary Science Stage 1
RESEARCH	Constructing research questions • Ask basic questions about a given topic	• Topic 1.1 Question 3: SB p3, TG p3 *Elicit and help learners form questions about the plants.*
	Information skills • Talk about information on a given topic in sources provided	• Topic 5.3 Activity 3: SB p81, TG p81 *Ask learners to display their fact sheets and talk about them.* • Topic 6.2 Activity 3: SB p87, TG p87 *Ask the groups to display their posters and introduce the information.* • Topic 6.5 Activity 3: SB p93, TG p93 *Ask learners to display their fact sheets and talk about them.*
	Conducting research • Begin to participate in simple investigations and ask basic questions to find information and opinions	There are many simple investigations provided which give learners initial experience of conducting research. For each, focus them on some questions about the topic and encourage them to predict answers before they do the investigations. Some examples are: • Topic 2.7 Activity 3: SB p29, WB p22, TG p29 • Topic 3.2 Activity 3: SB p35, WB p26, TG p35 • Topic 3.6 Activity 3: SB p45, TG p45 • Topic 3.8 Activity 3: SB p47, TG p47, PCM C6 • Topic 3.10 Activity 3: SB p51, WB p45, TG p51 • Topic 3.11 Activity 3: SB p53, TG p53, PCM C7 • Topic 4.1 Activity 2 & 3: SB p57, WB p49 & 50, TG p57 • Topic 4.5 Activity 1: SB p65, WB p57, TG p65 • Topic 6.1 Activity 3: SB p85, WB p79, TG p85
	Recording findings • Record information on a given topic in pictograms or simple graphic organisers	• Topic 3.1 Activity 2: SB p33, WB p26, TG p33 *Introduce and demonstrate the use of Venn diagrams. Talk about how they are useful to record information.*

Cambridge Global Perspectives	Learning Objectives for Stage 1	Collins International Primary Science Stage 1
ANALYSIS	Identifying perspectives • Say something known about a topic	• Topic 1.3 Activity 3: SB p7, TG p7 Learners talk about the differences between animals and plants using their drawings for focus. • Topic 3.3 Activity 3: SB p37, WB p28, TG p37 When they have completed WB p28, ask learners to hold up one of their chosen objects and describe it. • Topic 3.11 Activity 2: SB p53, WB p47, TG p53 When they have completed WB p47, ask learners to talk about one of their objects saying what it does and the materials it is made of. The **How well do you remember?** sections in the **Looking back** units are all suitable opportunities for learners to talk about what they have learned. • SB p14, TG p14 • SB p30, TG p30 • SB p54, TG p54 • SB p74, TG p74 • SB p82, TG p82 • SB p94, TG p94
REFLECTION	Personal viewpoints • Talk about what has been learned during an activity with support	• Topic 1.6 Consolidate and review: WB p11, TG p13 Learners talk about what they have learned in pairs using their completed WB p11 activity as support. Elicit some point for general discussion. • Topic 2.4 Consolidate and review: TG p23 Learners discuss what they have learned as a class and then in groups. Put key words on the board as support. • Topic 3.8 Activity 2: SB p47, WB p40, TG p47 Focus on Q4 on WB p40 and hold a class discussion. • Topic 4.4 Activity 2: SB p63, TG p63 Focus on the discussion after completing the constructions and encourage learners to express what they have learned and could do to improve their waterwheels.
COLLABORATION	Cooperation and interdependence • Share resources with others while working independently or with a partner	There are a number of activities for which learners will be provided materials that they need to share in order that they can all complete the activity. Monitor this and guide learners to collaborate. Some examples are: • Topic 3.3 Activity 3: SB p37, TG p37 • Topic 5.2 Activity 2: SB p79, WB p75, TG p79 • Topic 6.4 Activity 2: SB p91, TG p91
	Engaging in teamwork • Work positively with others	There are many group activities in which learners need to work as a team to complete an investigation or task. Monitor these and give guidance where it not happening. Some examples are: • Topic 1.6 Activity 1: SB p13, TG p13 • Topic 3.5 Activity 3: SB p41, TG p41 • Topic 3.7 Activity 3: SB p45, TG p45 • Topic 3.8 Activity 3: SB p47, TG p47 • Topic 4.4 Activity 2: SB p63, TG p63 • Topic 4.5 Activity 2: SB p65, TG p65 • Topic 6.3 Activities 1 & 2: SB p89, TG p89

Cambridge Global Perspectives	Learning Objectives for Stage 1	Collins International Primary Science Stage 1
COMMUNICATION	Communicating information • Answer questions with relevant information about a given topic	*Many of the **Consolidate and review** sections at the end of each Topic in the TG notes are suitable opportunities. Some examples are:* • TG p7 • TG p9 • TG p29 • TG p69